The Giving Tree...
A Metaphor for Climate Change

by

Dr. Mark J Poznansky and Shoshana Israel

The contents of this work, including, but not limited to, the accuracy of events, people, and places depicted; opinions expressed; permission to use previously published materials included; and any advice given or actions advocated are solely the responsibility of the author, who assumes all liability for said work and indemnifies the publisher against any claims stemming from publication of the work.

All Rights Reserved
Copyright © 2023 by Dr. Mark J Poznansky and Shoshana Israel

No part of this book may be reproduced or transmitted, downloaded, distributed, reverse engineered, or stored in or introduced into any information storage and retrieval system, in any form or by any means, including photocopying and recording, whether electronic or mechanical, now known or hereinafter invented without permission in writing from the publisher.

Dorrance Publishing Co
585 Alpha Drive
Pittsburgh, PA 15238
Visit our website at *www.dorrancebookstore.com*

ISBN: 979-8-8868-3113-9
eISBN: 979-8-8868-3972-2

FIG. 1 THE BOY AND THE TREE

The Giving Tree *is a children's book written by Shel Silverstein in 1964. But this is not a children's book. It is an accounting of how we, in our relationship with the tree and by extension coal and other fossil fuels, have changed the Earth's climate through global warming. We discuss what the implications are for our very existence, and what we must do to reverse those changes in order to survive and, in the extreme, avoid extinction. We wrote this book because we have grown to appreciate the words of the young Greta Thunberg to the United Nations when she said:*

"You say you hear us and that you understand the urgency. But no matter how sad and angry I am, I do not want to believe that. Because if you really understood the situation and kept on failing to act, then you would be evil. And that I refuse to believe."

She went on to say:

"You have stolen my dreams and my childhood with your empty words. But I don't want your hope. I don't want you to be hopeful, I want you to panic. I want you to feel the fear I feel every day and then I want you to act."

This work is written to give you a better understanding of what that existential crisis really is, to help you understand why Greta Thunberg is panicking the way she is, and perhaps most of all, to help you understand why she is correct to panic.

DEDICATION

In a very real sense, this book is dedicated to the future of humanity: to our children and their children, and to all the generations that follow.

It is specifically dedicated to our grandchildren/children: Avi, Effi, Olivia, Gil, Zachary, Rafi, Zev and Orly.

The book is written with the urgent warning that in order for them to live healthy and prosperous lives even remotely resembling our own, we must commit ourselves to protecting our environment and solving the issues of global warming and climate change.

CONTENTS

CHAPTER 1: INTRODUCTION 1

CHAPTER 2: THE TREE'S STORY 13

CHAPTER 3: FIRE and COAL, OIL and GAS 19

CHAPTER 4: THE BOY'S STORY 25

CHAPTER 5: A BRIEF INTERLUDE in ART 31

CHAPTER 6: THE ROLE TREES PLAY 35

CHAPTER 7: METHANE'S ROLE 55

CHAPTER 8: THE RELATIONSHIP SOURS 61

CHAPTER 9: THE ANTHROPOCENE 65

CHAPTER 10: DEFINING GLOBAL WARMING 71

CHAPTER 11: FACT CHECKING the CLIMATE CHANGE DENIERS 89

CHAPTER 12: FIRST CAME DEFORESTATION,
    THEN COMES REFORESTATION 95

CHAPTER 13: CLIMATE CHANGE and the BOY'S HEALTH 109

CHAPTER 14: THE COSTS OF CLIMATE CHANGE 117

CHAPTER 15: GLOBAL WARMING IS A FACT 123

CHAPTER 16: DEFINING THE CRISIS 145

CHAPTER 17: THERE ARE SOLUTIONS 157

CHAPTER 18: R.I.P. (REST IN PEACE) OR REQUIEM FOR A TREE 187

APPENDIX: A BRIEF HANDBOOK FOR DOING COMBAT
    WITH A CLIMATE CHANGE DENIER 197

SOURCES and ADDITIONAL READING 211

Chapter 1
INTRODUCTION

I am a tree. Well, that's not entirely true. I am a tree representative of all trees: past and present, and with some good fortune into the future as well. And I'm about to tell you my story, a story that ranges over a period of more than 500 million years. In fact, the story can even include my earlier ancestors, the first "green things," like mosses, algae, and very simple plants. These have existed for twice that amount of time, 1 billion years, and maybe even longer.

And I am a boy, and I represent…well, really all of humanity. Well, maybe not all humanity, but at least that which is commonly known as "Modern Man." My story is at least 1,000 times shorter than that of my good friend the tree; maybe only 300,000 years or so. Some say that I'm really older than that since there is evidence of something called Homo Erectus who appeared 1.5 million years ago, and four legged primates that go back even further (40-50 million years). But 300,000 years is the age given for "Modern Man," so let's stick with it.

My relationship with the tree was, for the most part, very positive. Between roughly 1 million years BCE (which was well before I even became "Modern Man") and 1950 CE, which is pretty recent, we got along really well. There were no major problems. Perhaps maybe just a few… you might call them skirmishes.

I cut down a bunch of trees to clear land to farm, for wood to heat my house and even, in fact, to build it. There were lots of trees, and nobody really noticed.

But by 1950 or thereabouts, the relationship started to sour, and it's gotten worse year by year. Today it's a real problem. Some even say that it's put all of humanity—that's of course all of us—in grave danger. A few even say that we humans may even be in danger of becoming extinct. Some use the expression that our relationship has now become an **"existential threat."**

To tell you the truth, it's entirely my fault. It's also becoming very clear that if we (e.g., the boys—and girls) are to survive even just another hundred years or so, we're going to have to make some pretty serious changes. It's not a matter of, "Sorry, we'll try not to do it again." We're going to have to dramatically change the way we do things. We're going to have to change how we treat many aspects of our surroundings, especially our environment. In particular, we're going to have to change how we interact with trees, both alive and dead, who were for so many years our good buddies. I'm talking of course of how and where we source our supplies of energy and how we continue to cut down trees.

(Just a note about nomenclature to make sure that we're all on the same page: BCE means "before the common era." It used to mean "before Christ" (BC) until we became religiously and politically correct. And of course CE refers to the "common era," or after year zero, the day given for the birth of Christ, which is assigned as the start of the common era. When we talk about temperatures, we're going to be using the international or more commonly accepted Celsius or metric scale. If you'd like to convert, one degree Celsius is 1.8 times larger than one degree Fahrenheit).

But let's go back and rediscover the history of the tree and the history of the boy. We'll start with the tree, because after all, she's a lot bigger and a lot older. Let's recount a bit of her history.

It's not exactly clear when I (the tree) first arrived, but it was under 500 million years ago, so I'm nowhere near as old as some stuff…but I am a lot older than you. And in case you were wondering, here's a definition of who and what I am: I'm a perennial plant with a long stem, usually called a trunk. I come in many different shapes, sizes, and even colors, and I possess the secret of longevity. I have been known to live for thousands of years. When I was last counted, just a few years ago, there were more than 3 trillion mature trees in the world. But that number has dropped by almost 4,000 trees since you started reading that last sentence. Can you imagine? That's 700,000 trees disappearing each and every hour of the day! Think about it for a minute. That is the number of trees being cut down per hour not per day, month, or year. Some have taken to calling it systematic deforestation on a massive scale and that's an important part of my history. The rate at which I'm disappearing, or really being cut down, is impressive, and it's a real problem, not only for me but for you as well. And that's why I'm telling this story. You may consider yourself an ordinary person, a "Johnny-come-lately," if you will but your impact on me and my friends has been huge, and not all good. And that's the story that I'm about to tell.

Since it's not just about me, to jog your memory, I've compiled a brief history of the Earth, at least in terms of its various lifeforms and my many ancestors, who, as a matter of fact, also contribute to the "problem," though not strictly speaking on their own.

FIRST APPEARANCES
(These are rough approximations, but you'll get the drift).

3.7 billion years ago:	Simple single celled organisms, bacteria and microbes first appear, some 800 million years after the Earth was formed.
3.5 billion years ago:	Here is the first evidence of photosynthesis (defined below) and the production of oxygen on Earth…but in very miniscule quantities.

1.7 billion years ago:	There is the first appearance of multi-cellular organisms and sex…the earliest evidence of sexual interactions between living entities.
850 million years ago:	The first complex land plants start to appear on open, barren lands.
530 million years ago:	First fish appear in the oceans during what is called the Cambrian explosion, which was a huge expansion in the variety of lifeforms on Earth.
370 million years ago:	I ARRIVE, THE FIRST "REAL" TREE WITH A TRUNK, REACHING TOWARDS THE SKY.

While I will admit blame for some of the negative occurrences on Earth today, I also take credit for providing an atmosphere where all living animals can live and thrive. You see, before I, and my other green predecessors, came along, there was little or no oxygen in the atmosphere, so you wouldn't have been able to breathe. The earliest green plants (which pre-date me by approximately 500 million years) started to leak oxygen into the atmosphere through the process of photosynthesis (you'll see the definition shortly). By the middle of the Devonian Period (400 million years ago), green plants and early trees (originally our friends the Gilboa trees) were spewing out enough oxygen and producing enough soil and shelter for small animals to thrive. That was our role in altering the pace of change on Earth. By spewing out all that oxygen, we sped up the process of evolution and allowed for whole new varieties of species to evolve.

ANOTHER BUNCH OF FIRSTS

350 million years ago:	The earliest land vertebrates appear arriving from the sea.
200 million years ago:	We see the first appearance of the dinosaurs.
60 million years ago:	The first primate appears.
7 million years ago:	The earliest apes appear.
350,000 years ago:	The boy, the first Modern Man, has arrived.

And so, just to put it in the context of our own (the boy's) development, here is another short **LIST OF FIRSTS**. It's pretty impressive to see just how young we really are compared to earlier life forms.

4,000,000 years ago:	Earliest bipedal animals arrive (standing on two feet).
350,000 years ago:	First appearance of Modern Man.
320,000 years ago:	First appearance of stone tools.
150,000 years ago:	First use of fire.
75,000 years ago:	First evidence of cooking.
50,000 years ago:	First evidence of clothing.
20,000 years ago:	First use of organized speech.
10,000 years ago:	First human settlements.

So now, let's get back to me, the tree, and have a look into how I came to be. I come from a pretty humble background. Me and my cousins, there were many of us, started off as small, fern-like plants growing right on the land. But we were known to be relatively competitive, and since we all need energy from the sun to grow, we tried to outgrow one another by reaching up to the sky in order to capture those sun rays and grow to be taller and stronger than our fellow cousins or friends. And as we grew taller, we had to grow stronger stalks or stems so that we could grow higher to capture more of the sunlight without literally falling down. With such growth, we then developed wood to keep our stems strong so that we would not be brought down by winds or storms, or by our competitors.

How did we get to be so strong? Well, it starts with a simple chemical process which is pretty much fundamental for life on Earth, at least the lives of all plants and animals and that's called:

PHOTOSYNTHESIS

The process of photosynthesis is basic to the vast majority of life forms on Earth. It is the way by which green plants and certain other organisms actually transform the light/radiant energy from the sun and convert it into chemical energy stored in the plant. The wood of a tree is perhaps the best example and the basis for much of this story. That stored energy is of course the same energy that we ultimately use for almost everything we do (heat our homes, grow our food, drive most of our cars, and of course power all our industries). Chlorophyl, the green stuff, has to be present to act as a catalyst.

The chemical reaction is simple enough:

$$Carbon\ Dioxide + Water + Sunlight \xrightarrow{Chlorophyl} Sugar + Oxygen$$

Or more scientifically-speaking, using chemical terms:

$$6CO_2 + 6H_2O + Solar\ Energy \xrightarrow{Chlorophyl} C_6H_{12}O_6 + 6O_2$$

Which translates to six molecules of carbon dioxide plus six molecules of water, add some solar energy in the form sunshine, and in the presence of a green leafy plant, which contains chlorophyl, you get one molecule of glucose and six molecules of oxygen. The glucose molecule that is produced contains six atoms of carbon, 12 atoms of hydrogen, and six atoms of oxygen. As an added bonus, but one that is critical to human and all animal forms on Earth, the reaction also produces six molecules of breathable oxygen released into the air, which is of course what we require to live. And that's all the chemistry that you'll need to understand our relationship.

It's really pretty simple, then I, the tree or plant, take all the sugar molecules, technically it's glucose, and link them together to make more complex sugars like cellulose, which is the most important component of wood. Put simply, the energy from the sun is captured and stored directly in complex sugar structures in wood, giving it its strength. This is essential in understanding what fossil fuels are really about, which is a critical part of my ongoing story. The energy is stored in the chemical bonds between the carbon (C), hydrogen (H), and oxygen (O) atoms. The energy remains in those bonds until it is released, usually by combustion or burning. Here's a rough analogy: Take a rubber band

and stretch it. You've now taken energy from your hand and arm muscles and transferred it to the elastic components in the rubber band. If you do nothing, the energy remains in the rubber band, but if you let go of the rubber band, the energy is released as kinetic energy, and the rubber band flies across the room…and might even break something. Or, if you wish, you can keep the elastic band stretched and "store" that energy until you want to use it. That's what happens to the energy stored in fallen trees or other vegetation that die.

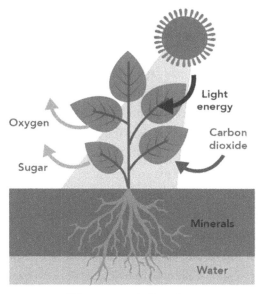

FIG. 2 HOW PHOTOSYNTHESIS WORKS

So, what did we (the trees) do for a living other than grow higher and higher and compete for the sunlight with our cousins? Well, in addition to nurturing ourselves, we provided sustenance (food and shelter) for a whole host of different lifeforms, probably numbering in the hundreds of millions of different species. At first, we worked primarily with tiny microbes maybe 1-2 billion years ago when we were just plants, and then we went on to provide comfort, food, and shelter to simple insects, which appeared 400-500 million years ago.

By the time the first birds arrived seeking shelter and food around 60 million years ago, we had matured a lot, and we were starting to look like the trees that are seen in some forests today. Those many critters ate some of my green leaves, and for the most part, it was fine, as we had the wherewithal to regrow the vegetation through that process I just described "photosynthesis," getting energy from the sun, water from the Earth, and carbon dioxide picked up out of the atmosphere. And when I think about it, we had a pretty good relationship with those living creatures. I took the energy from the sun, and I used it for both my own growth and sustenance, as well as theirs. And of course, I also gave off oxygen for many species to live and "breathe."

Through the course of time, many of us died from any number of causes. Some of us simply died of old age; some of us didn't get enough sun or water, or weather did us in. Some of us even got infected by some bug or disease that we had no defense against, and we died much earlier than expected.

And so, after a time, either by natural incident, by bad chance, or simply by old age, I died. I didn't have a "proper burial." In fact, I just lay where I fell, and then other close and not so close relatives also died, and they just fell on top of me. And so generation after generation continued to grow and then die and lay to rest on top of one another, and of course my own remains. With the accumulating weight of all these remains, we sank deeper and deeper below the Earth's surface. Remember we're talking about millions and even hundreds of millions of years of life and death. Key to our story is that all that energy I had stored as carbon products in the wood didn't disappear; they just remained stored in my "remains" deeper and deeper below the Earth's surface. If you were somehow floating in space above the Earth and observing what was going on down there, you would probably think that all of that energy had just disappeared as layers upon layers of material, most often vegetation such as plants and trees, and all manner of other living things continue to grow accumulate energy, die, and be buried. And that would be true for many hundreds of millions and even billions of years.

The story could have just ended there, but it didn't. In parallel, or really a lot later on, as recently as 500,000 years ago, a new lifeform appeared on Earth, an earlier relative of Modern Man and some of his/her immediate ancestors, and they made a critical discovery. They discovered a hard, black substance, coal, either in caves or just below the Earth's surface. They discovered that the substance, when burning, would give off a tremendous amount of heat. They had of course discovered the energy contained in fossil fuels. Many years later, other remains of trees, vegetation, and other life forms would be found deep in the ground in the forms of oil and gas, which would become even more important to our story. Today it is of course humanity's primary source of energy to heat and light homes, drive cars and industries, and provide food. Now these folks, even some 300,000 years ago, were a pretty inventive group and realized that they could make good use of that stored energy for a whole variety of reasons, but initially, it was for heating and cooking. And that also is the main story of this book. But more about that later.

So, what is the flip side to photosynthesis? What happens when we actually take that coal (or it could be oil or gas…collectively called "fossil fuels") and burn it for heating purposes or to provide us with energy that humans seem to need for almost anything they do?

The chemical reaction is the exact opposite of photosynthesis. And this is the release of the energy that originally came from the sun that has been stored in the carbon bonds these many years in the remains of trees and other plants. These are the fossil fuels. The fuels are burned/ignited, and the energy that is released is used to heat our homes, drive our cars, fly our planes, and the list is obviously endless.

$$C \text{ (complex sugars in wood)} + O_2 \xrightarrow{\text{Burn}} CO_2 + Heat/Energy$$

On the surface, burning coal is quite simple. The coal is heated until it reaches its combustion temperature, at which time it starts to burn. The carbon in the coal combines with the oxygen in the air to create carbon dioxide and release energy in the form of light and heat, the same solar energy that the plant/tree used in the first place to grow through the process of photosynthesis. It is the released carbon dioxide that results from burning fossil fuels that becomes the major culprit in global warming and climate change, and we'll elaborate on that at length in the coming chapters. You might be wondering, and you should be, why we can't use the energy from the sun directly and not have to go through that whole photosynthesis, life, death, fossil fuel cycle to collect and use that energy. And you would be right to think about it, and we'll talk a lot about that in a later chapter. We'll also discuss how that direct energy from the sun can and should be used and also about why it isn't…yet. That's of course the basis of all our current talk about solar panels, solar energy, and new types of batteries.

But not all coal is created equally. It depends on who it slept with these past millions or hundreds of millions of years, where it's been, what it's picked up over its buried lifetime, and of course how the world chooses to extract it from deep within; to use it, and to burn it. There are some pretty nasty things that can happen in burning coal, especially if it's what's often referred to as "dirty coal." If it happens that there is not enough oxygen in the air when the coal is ignited, the reaction can still take place, but it can produce CO (carbon monoxide) rather than CO_2. Carbon monoxide, unlike carbon dioxide, is a very toxic molecule, especially in humans, where it can displace oxygen on the hemoglobin molecules in our blood, thereby starving our cells of oxygen and quickly resulting in death. This used to be all too common in homes with inefficient or poorly maintained heating systems. And this is why it is essential to have carbon monoxide detectors in any setting where coal, oil or gas, or even wood is being burned.

Coal also often contains very substantial levels of nitrogen and sulphur products, and when it is burned, it gives off toxic molecules like sulphur dioxide (SO_2) and nitrous oxide (N_2O). Nitrous oxide, like carbon dioxide, is also an important greenhouse gas, which when it accumulates in the atmosphere can be a significant contributor to global warming. There's another negative aspect about burning coal and especially dirty coal. When SO_2 and NO_2 are released into the atmosphere and get dissolved in rainwater, the resulting acid rain can be harmful to all matter of crops and soil, creating damage to buildings, and of course spoiling the air we breathe. It also leads to the acidification of our oceans, lakes, and rivers with frightening impacts on virtually all matter of marine life. As early as the 1960s, when coal was in its heyday , there were lakes and seas that were declared officially dead as a result of acid rain. In some instances, the extreme acidification of their waters resulted in the killing of most, if not all, of the marine life. We've also all heard about how so many of the coral reefs in oceans around the world have been destroyed through acidification of the seas. This is all a direct result of our use of fossil fuels, primarily coal, to provide us with the energy we "need." It's pretty embarrassing to realize just how much damage we've done.

Chapter 2
THE TREE'S STORY

So now that I've given you a brief overview, let's go back and fill in the blanks and allow you to delve a little deeper into my background and that of my many cousins. I'm not only going to give you my own (the tree's) history, but I'm going to include all matter of plants and even a couple of animals as well.

5 billion years BCE:	Following the explosion of a huge star, Earth is formed as a result of the condensation of a ball of hot gases. By far, the most common gas in the atmosphere was then CO_2.
3.4 billion years BCE:	Ancient rocks reveal the first evidence of photosynthesis, resulting in the production of the first minute amounts of oxygen (from carbon dioxide) in the oceans.
1.6 billion years BCE:	The earliest plants (in the strictest sense a microbe), green algae, are known to have lived within the Earth's oceans. The land consisted of barren rock with a sparse distribution of bacteria

and fungi. With time, atmospheric carbon dioxide begins to fall, at first just a bit, and at the same time, small amounts of oxygen appear in the atmosphere.

500 million years BCE: Plants are largely small and marine-based, but increasingly some green algae and other microorganisms are adapting to grow on land. The high carbon dioxide in the atmosphere is the cause for the high temperatures on Earth during what is called the Precambrian Era. But the terrestrial plants, by utilizing carbon dioxide to grow, are ever so slowly depleting greenhouse gases in the atmosphere, causing the Earth to cool and providing oxygen for other life on land to flourish, sort of a prelude of what's to come.

400 million years BCE: This is sometimes referred to as the period of the Cambrian explosion. There is a huge expansion of "life" on earth. Diverse simple plants can be found adapting to living on land and thriving as they proceed to increase the amount of oxygen in the atmosphere. This also decreases the amount of carbon dioxide (remember photosynthesis) in the atmosphere, and cooling occurs.

350 million years BCE: During this, the Devonian Period, soil has begun to appear on the land, and the plants have developed circulatory systems, allowing them to "stand upright" and draw water from the land. As the plants grow stems, evidence of wood structures begin to appear.

300 million years BCE: The general overall climate has become drier and somewhat cooler, and forests begin to appear in many of the more temperate regions of the Earth. During this, the Triassic Period, the first dinosaurs appear, having evolved from reptiles.

200 million years BCE: This is the Jurassic Period, covering 56 million years. The climate has become wetter, leading to the development of wetlands and jungles. It is also a very rich period in animal development as birds evolve from a branch of therapod dinosaurs.

100 million years BCE: The climate starts to dry, and extensive grasslands join with forested areas to provide food for large grazing mammals as well as for the protection of smaller animals, including rodents.

1 million years BCE: As the climate cools, large tropical forests die off and even larger grassland areas appear. There are many hundreds of different species of primates living across many parts of the Earth.

300,000 years BCE: I meet my first Modern Man. By 13,000 BCE, there is plenty of evidence of humans living as hunter-gatherers, living in communities, and cultivating plants as sources of food. Think about how long this took to happen. This is also the period of multiple ice ages. It has been suggested that during one of the coldest periods of the last ice age, some 20,000 years ago, prehistoric hunter gatherers deliberately set forest fires to create areas where food gathering would be easier. This is likely why large parts of Europe are less densely forested. It's a complex time with lots of vegetation, lots of varied animals, and humans starting to get to work. It is also a time where climates fluctuates over long periods of time as, for example, in the coming and going of ice ages. But note that these changes occurred over tens of thousands of years and are generally explained by the position and tilt of the Earth relative to the sun. As we'll see and prove later on in our story, this is very different from the man-made and very rapid changes that our climate has been experiencing just over the past 20 years or so.

2020s CE: That's now, and things are in incredible flux, created much more so by the tremendous population explosion and the activities of man rather than the traditional/previous forces of nature. For example, approximately 370,000 trees were cut down in the last 30 minutes since you started reading this book. That's 18 million (18,000,000) trees per day or 6.5 billion (6,500,000,000) trees per year. If you just happened to look down at Earth from the international space station, it would immediately occur to you just how much of the Earth has been transformed from massive areas of grasslands and forests to concrete jungles along with complex networks of highways and highly cultivated agricultural lands. The changes have been extraordinary and have occurred in such a short period of time with no evidence of slow-down in sight. There doesn't seem to be any plan in place, and if you happen to wonder whether this pace is at all sustainable, it isn't.

Read again: 6.5 billion trees are being cut down each and every year.

Authors note: Every time we see or write down that number, we feel we have to go back to our sources and check the number again. But there it is; within a +/- 10 percent error, that is the correct number. Quite incredible.

Chapter 3
FIRE and COAL, OIL and GAS

FIG. 3 FROM TREES to COAL to OIL

It took me (this is the tree speaking) many years, about 370 million to be more precise, before I began to realize what a profound negative impact I was to have on the Earth and specifically on many of its lifeforms, including humans. This has become especially clear over the past 70 years (1950-2020) following the economic and industrial boom that followed World War II. The year 1950 is often used as the year of transition when things really began to change. But when I look at it more critically, it started somewhere at the beginning of the Second Industrial Revolution around 1850 and probably coincided not only with the huge population growth, but also with the development of the steam engine. By the year 2000 CE, it had become absolutely clear that I was playing an important role in putting all of humanity (as well as other life forms) in grave danger. In fact, many have concluded, and I can't disagree, that by this time, I was playing the critical role. I had become responsible for exposing all of humanity to a very major crisis and that of course is global warming. Some have even suggested that many lifeforms on Earth, especially the boys and the trees, were facing a major existential crisis—as in their very survival—even in their destruction and extinction. It's just that serious.

Wow, that sounds like a serious guilt trip if there ever was one! How did that even happen? My ancestors and of course my contemporaries knew about fire. They were aware of the sometimes devastating impact that fires caused by lightning and rupturing volcanoes were having on trees and forests and other vegetation, and many other lifeforms as well. At some point around 1 million years ago, an early version of Modern Man, Homo Erectus, is thought to have begun the first routine "controlled" use of fire to cook food, provide warmth, and even forge certain types of tools. They probably learned, quite by accident, from the impact of uncontrolled fires and the release of heat from the burning material. With time, the boy was routinely using dead or even live trees for his various purposes. I wasn't thrilled to have those humanoids use me as a source of firewood, but in the entire scheme of things, it wasn't that serious. For example, as of around the year 50,000 BCE, there were trillions of trees, probably 2 or 3 trillion anyways, and very few, maybe less than 100,000 of them (i.e., humans), so maybe 30 million trees per human. We, the trees, didn't really have that much to worry about. And for the most part, the humans were just burning wood, and most of that would have been from fallen or even dead comrades of mine. Not a very big deal.

The Giving Tree…A Metaphor for Climate Change

As recently as 3,000 years ago, the Chinese are reported to have begun to routinely collect (i.e., mine) and use a shiny black material as a source of warmth and light. That shiny black stuff was of course coal, a remnant of my long-deceased ancestors. The "so-called" caveman, who first discovered coal, immediately realized that it was better to burn coal rather than wood because it burned hotter and longer, and they wouldn't have to scavenge for wood to burn as often. They certainly wouldn't have expressed it in these terms, but they had discovered that a piece of coal would release more heat energy than a similarly sized piece of wood.

I must tell you that I'm really not that crazy about talking about coal, or for that matter oil and gas. First of all, the "stuff" that we're talking about are my ancestors. These are my folks—well, not really folks, but trees and all matter of vegetation and other life forms who went to their graves long ago. But now, even worse, they have been coming back in recent years to haunt us and wreak havoc on our climate and threaten our way of life and maybe even our very existence. While that may sound just a tad dramatic, it's not such a bad analogy. And note it's not just the boy's existence that we're talking and worried about, but mine as well.

So what is coal, really? It's called a fossil fuel. Wikipedia defined a fossil as "any preserved remains, impression, or trace of any once-living thing from a past geological era." Examples include bones, shells, exoskeletons, and of course wood and any and all other matter of vegetative material that has died and been buried thousands and millions of years ago. Coal was formed over long periods of time. Some coal has been dated (yes, by carbon dating) to be older than 650 million years.

Carbon dating is a technique used to determine the age of many different objects, and it's done by determining how much decay has occurred of a natural radioactive substance. In this case, it's the radioactive form of carbon or carbon-14 in the material whose "age" is being determined. At a time even before the age of dinosaurs, large parts of the Earth were covered with swampy lands where all sorts of ferny plants, my ancestors, and other forms of vegetation grew. As the plants died, they would sink to the bottom of the swamps. This happened repeatedly over hundreds of millions of years, and the thick layers of dead plant material were packed down hundreds and thousands of feet deep. With the huge pressures from the material on top, earth, dirt, and other dead vegetation, the material was compacted and formed what we know as coal. Think of it for a moment. All that energy from the sun and photosynthesis that my ancestors accumulated was now "packaged" in this very dense, black material. Note the fact that the coal was formed over a long period of time, and once used, it is no more. That's why we call this form of fossil fuel a non-renewable resource. While this is true and the use of fossil fuels as an energy source might be considered non-renewable, the sheer quantities of these fossil fuels are almost unfathomable, and for all intents and purposes, the supply may be considered unlimited. And that, as we'll see, is not such a good thing.

Coal deposits were discovered in England in the early 1700s, and the hot burning fuel was used as a source of heat and a way to cook food. As long ago as the 1300s, the Hopi Indians in the western United States used coal not only for cooking and heating but also to bake the clay that they used to make pottery.

It was really during the heydays of the Second Industrial Revolution that coal really came into its own, as huge amounts of energy were required to fuel the steam engines that drove the revolution and then the blast furnaces, which were used to forge steel. It was also around this time in Britain and a little later in the United States, after 1850, that enormous deposits of coal were discovered in the north of England and in the eastern United States of Virginia and Pennsylvania. The fact is that coal is ubiquitous and has been mined in every single state in America and around the world. The discovery and use of coal was incredibly important, and it brought huge benefits. But it was not without its problems. We have only to remember the black faces of the coal miners coming out of the deep mines in West Virginia and in the north of England, and the horrible diseases that many of the miners succumbed to, one of the most common being known as black lung disease.

Oil and gas have almost identical origins to coal in that they are also fossil fuels derived from the remains of prehistoric (i.e., many millions of years ago) plants and animals. The differences between coal, oil, and gas are quite simple. On a superficial level, you might simply say that the three are really the same: coal as a solid, crude oil as a fluid, and natural gas as a gas; and that's pretty close.

Coal is a combustible black sedimentary rock comprised primarily of carbon that was formed when dead plant matter, including especially trees, decayed into peat. Over millions of years, the heat and deep pressure converted the peat into coal. Crude oil (also sometimes called "petroleum" in its initial form) is produced when large quantities of dead organisms (plant, animal, and microbial organic material) are buried beneath sedimentary rock and subjected to both intense heat and pressure and exist in varying liquid forms. You might say that oil is simply a liquefied (under pressure) form of coal. Natural gas (also called fossil gas) is formed from the same buried organic material that produces coal and oil, only it is under even greater pressure and exists as a gas or vapor. But it can also be formed quickly, and its low density allows it to rise to the surface, as in gases coming out of landfills. Natural gas is sometimes considered "cleaner" than oil because it is highly combustible, burns more efficiently, and releases less sulphur dioxide and nitrogen oxide into the atmosphere. But it's hardly clean in terms of how much carbon dioxide is produced when it's burned.

So let's not forget that with these various fossil fuels, the primary product in each case is carbon dioxide, currently the most abundant and concerning of the greenhouse gases that are responsible for global warming. We'll learn about the sources and concern about greenhouse gases (GHG) in subsequent chapters.

Note that all three of the forms of fossil fuels can come in various degrees of purity. That is, they may also contain various amounts of nitrogen, sulphur, phosphate, and other elements. The degree of contamination of the carbon-based fossil fuels with these compounds determines whether the fuel is considered to be dirty or clean.

Chapter 4

THE BOY'S STORY

Now, let's revert to my story, as it were; the life of the boy. How did I get to where I am? How have I changed over time? And most important, what has been my role in global warming and climate change? We aren't going to take this opportunity to offer a lesson on evolution (we'll leave that to Darwin), this is a story of change: enormous and, recently, extremely rapid change.

Earlier I said something about having been around for just 200,000 to 300,000 years. That's true for the modern species of human beings, or "Modern Man," Homo Sapiens who emerged out of Africa in that period, spread, and expanded to other continents.

FIG 4 THE BOY AND THE SMOKESTACK…A PROBLEM

But the history is much richer than that, so let's take a journey through time.

55-60 million years BCE:	This is the period when the very earliest primates appear. Primates are a biological order containing all species surprisingly related to lemurs (a small animal with a pointed snout, large eyes, and a long tail), and of course monkeys and apes, with the last category eventually including humans
5-8 million years BCE:	This is the period when two species diverged to create separate lineages, one evolving into gorillas and chimps, and the second into the ancestors of early humans called "hominids."
4 million years BCE:	Homo Erectus (which literally means "upright man") first appears. Evidence for these earliest bipedal and upright walkers was found in Africa.
2.6 million years BCE:	Evidence of early stone toolmaking appears with a chopping tool used by an early human ancestor, the Homo Habilis (sometimes called "Stone Age man"). They were found in Ethiopia and other areas of eastern and southern Africa.
1.8 million years BCE:	Homo Erectus expands out of Africa with fossil evidence appearing in Europe and in China.
1.0 million years BCE:	Evidence of Homo Erectus first using fire for cooking and warmth.
800,000-400,000 years BCE:	During this period, fossil evidence reveals that the brains of Homo Erectus underwent signifi-

cant expansion. Apparently, we were now getting smarter.

400,000 years BCE: Neanderthals (Homo Neanderthalensis) first appear in various regions of Europe and Asia. Neanderthals were a distinct species of humans, who were thought to be mainly hunter-gatherers as opposed to Homo Sapiens, who had a more settled life producing food through agriculture and appearing to be more domesticated. The Neanderthals apparently survived through to about 40,000 years ago when they became extinct due to a variety of causes, including violence, disease, inability to adapt to climate change (the ice ages), and interbreeding with early Modern Man. And by the way, geneticists tell us that there is a little bit of Neanderthal in all of us. That's perhaps not really so surprising (sic), given some of our behaviour.

250,000 years BCE: Homo Sapiens (Modern Man) emerge out of Africa with an estimated population ranging from 10,000 to 30,000 individuals.

40,000 years BCE: Modern Man expands his territory from Africa to regions around the world—Europe, Asia, the Americas, and even Australia. Due to various climatic catastrophes, particularly a number of ice ages—the human population does not show significant growth during this period in history. However, there is evidence of culture (paintings on cave walls) and actual communities existing in this period.

10,000 years BCE:	There's been a huge population expansion around the world with the number of humans estimated to be anywhere from 1 million to 5 million. This is also the time of the retreat of the last ice age. There is evidence of early agriculture, the domestication of plants and animals occurring independently in at least 11 different areas of the Africa, Europe, Asia, and the Americas.
4,500 years BCE:	The world is getting more and more sophisticated. There is evidence of extensive communication through writing, and distinct civilizations have been developed, including actual organized state societies in Sumar in Persia (now Iran) and in Egypt.

And the rest, as they say, is history. The population grew slowly. Man had all sorts of challenges, not the least of which were weather and disease. It might surprise you that between the years 10,000 BCE and 2,000 BCE (that's a period of 8,000 years), the Earth's human population had risen from four million people to as many as 20 million. This is nothing compared to modern times when there are more than seven billion people, but a very substantial increase nevertheless. When looking at this timeframe, it's striking to realize how recent our presence on Earth is and then how quickly our numbers multiplied and how significant our footprint on Earth has rapidly become.

POPULATION GROWTH

Year	*Population*
10,000 BCE	4 million
2,000 BCE	20 million
Year 0	190 million

1700 CE	600 million*
1900 CE	1,650 million**
1950 CE	2,500 million
1975 CE	4,000 million
1999 CE	6,000 million
2020 CE	7,750 million***

* The Black Death/Bubonic Plague of the mid-fourteenth century killed 200 million people, about 30-50 percent of the world's population at the time.
** The Spanish Flu Pandemic of 1918-1919 killed at least 50 million people
*** The COVID-19 Pandemic of 2020-2022 killed more than 7.5 million people as of this book's publication.

A very important part of this story is the incredibly rapid expansion of the population on Earth, especially over the last 50 years. Picture this: *The population of humans on Earth has increased from 1975 to the present (just 46 years) as much as it did from the beginning of human life on Earth, say 400,000 years BCE until 1975.*

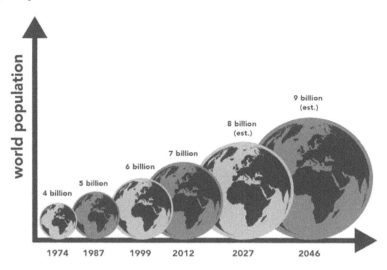

FIG 5 POPULATION GROWTH IN JUST 70 YEARS

The Earth is a very different place today compared to what it was just a hundred years ago. Men and women have not simply come forth and multiplied and built homes and cities and farms and factories. They have literally changed the face of the Earth in very substantial ways (remember that view from the International Space Station?). Some have even suggested that man has created a new geological age, the Anthropocene, an age in which human activity is having a profound impact on shaping the Earth in respect to its physical, geological, chemical, and biological diversity. And as we're beginning to learn, that may not be such a good thing. We'll cover that in greater detail in Chapters 8 and 9.

Chapter 5
A BRIEF INTERLUDE in ART

EMILY CARR and the Life of Trees

On the one hand, this is essentially a book about climate change. It is also a story of the "woes" that we've brought upon ourselves by our unique dependence on fossil fuels (and therefore the remnants of trees long gone) as our primary source of energy. We not only remember our dependent relationship with trees as Shel Silverstein describes in his wonderful children's book *The Giving Tree*, but also the essential nature of trees. Shel Silverstein's book was written in 1964, and while it's a wonderful book, it is also considered one of the most controversial of all children's books, the controversy is based on whether to consider the relationship between the boy and the tree as positive in terms of the tree giving everything of itself to the boy, or negative as in the boy is continuously abusive towards the tree. In either case there is no ignoring the magnificence of trees, or the splendor of Shel Silverstein's prose. In a beautiful book by Michael Jordan titled exactly that, *The Beauty of Trees*, the reader receives much insight into one of nature's most astonishing structures, their pure beauty and incredible diversity. It also describes the incredible intimacy that man has had with the tree not only in remote forests, but in our own backyards or next to our favourite park bench.

Any thought of man's relationship to trees immediately evokes one of the great Canadian Artists, Emily Carr. Though Emily Carr never became a member of the famous Canadian Group of Seven, one of their most famous artists, Lauren Harris, clearly referred to Emily Carr: "You are one of us".

Just shortly before her death, in frail health, Emily Carr insisted upon going out into the forest "to put on canvas those startling, vivid sweeps of swirling movement." These properties were so particular to her work and to the way in which she portrayed the trees and forests, and she is quoted as saying: **"I must go home today and go into the forest again. The forest still has something to say to me, and I must be there to hear."**

FIG 6 *TREES* BY EMILY CARR

Emily Carr did not only have a lifelong love and admiration of trees but clearly saw the dangers of unrestrained cutting down of trees. This was in the late nineteenth century in British Columbia. We'll cover in some detail in the next chapter why it is that man has become so quick and adept at cutting down trees.

Emily Carr gave the deforestation phenomenon a lot of thought, and she penned a political cartoon in 1905 with the caption. **"The Inartistic Alderman and the Realistic Nightmare,"** which included a short poem.

> "Ye ghosts of all the dear old trees, the oak,
> the elm, the ash, nightly those gentlemen go tease.
> Who hew you down like trash?"

For the rest of her life, she would rail against the logging industry's unrestrained tree cutting. Late in life, she admonished herself for not having done more to save the forests. It's not clear whether Emily Carr ever gave much thought about our reliance on fossil fuels as a primary source of energy. After all, it was only the early 1900s, but maybe, just maybe, she had a premonition. With some appropriate "justice," students at the Emily Carr University of Art & Design speak out loudly against the ills of the fossil fuel industry, requesting that Canadian Universities divest themselves from related industries. And they celebrate the tree in all its beauty, prominence, and importance.

It might be tempting to dream or perhaps to fantasize what would have happened if, from the start, we had all appreciated and adopted Emily Carr's attitude towards the forests. What would have happened if we had a greater respect for trees, both in their living forms as well as in their afterlife (i.e., fossils)? Perhaps today we would be using only renewable sources of energy and not be in the dire state of global warming and climate change. But unfortunately, that's not what happened – so the story must go on.

Chapter 6
THE ROLES TREES PLAY

"THE TREE GIVETH AND THE TREE TAKETH AWAY"

In Shel Silverstein's book *The Giving Tree*, from which part of our title is derived, we see the relationship of a single tree with a young boy. The tree gives shade and fruit and a place for a swing for our young boy to rest and play in the first part. Later, it gives wood for the young man to build a home, and finally the tree is reduced to a stump and only a place for the boy, now an old man, to sit. That story, in its simplicity, might be replicated for every boy (person) and every tree; it represents an ever repeating, renewable cycle in great harmony.

Unfortunately, the relationship has become anything but harmonious. In fact, it's probably pretty accurate to conclude that the relationship has now entered into a dangerous and destructive downhill spiral. The problem is quite simple: **There are simply too many people cutting down too many trees and burning too much fossil fuel.**

But it's not quite that simple. If we were to assume that deforestation, especially that which has been brought on by man's activities such as massive population increases and industrialization, is a relatively recent phenomenon, we would be wrong. Many, many forces over hundreds of millions of years have contributed to the health, or lack thereof, of the Earth's forests. In an earlier chapter, the Devonian Period, about 350 million years ago, was referenced, which represents a period of explosive growth of all manner of life, both in the seas and on land. And it was during this period that huge forest growth occurred and many different forms of animal life begin to appear on land.

It is not surprising and very germane to our story, that it was the evolution of the tree and of forests that was the major enabler of many of the animal life forms on Earth. Let's travel back a bit again and examine the role of plants, trees, and forests, especially as it pertains to the role of oxygen in evolution. But first, let's define some of the geological timeframes, for they are often referred to and are pertinent to our story.

GEOLOGICAL TIME FRAMES: A SIMPLIFIED VERSION

Precambrian: This is the time from the formation of the Earth about four to five billion years ago to the Cambrian "Explosion of Life" that occurred about 540 million years ago. The atmosphere was primarily made up of carbon dioxide with virtually no oxygen, and it was hot.

Cambrian Era: The time between 540 and 490 million years BCE saw a spectacular increase in the number of living organisms on Earth. This period is associated with an increase in atmospheric oxygen and an increase in temperature as glaciers retreated.

Paleozoic Era: This is a geological era starting around 540 million years BCE and ending 250 million years BCE. It started with the emergence of soft-shelled life in the sea and ended with the appearance of complex plants, insects, fish, and small reptiles. Sea levels were about 200 meters higher than they are today.

Devonian Period: This period lasted from 420 to 360 million years BCE. It was a period when large continents were formed and major plants and terrestrial animals emerged. Much of this development was arrested by a massive extinction event that occurred late in the period, destroying more than 80 percent of the known living species.

Jurassic Period: A geological time period between 200 and 145 million years BCE when dinosaurs were the dominant animal group.

Cretaceous Period: A geological time period from 145 to 65 million years BCE. It was the age of the reptiles, both on land and in the sea. It ended with another devastating mass extinction event, which probably occurred as a result of the Earth's collision with an asteroid or comet resulting in the killing off of all large animals (except birds) and many species of terrestrial animals.

Cenozoic Era: This is the geologic time period extending from 65 million years BCE and the mass extinction of non-avian dinosaurs to the present.

Paleogene Period: This is the geologic time period between 65 and 23 million years BCE. Many new animals, insects, and plants arrive, and the climate goes from hot and humid to cooler and drier. Note that these climate changes occurred over the course of tens of millions of years.

Neogene Period:	This geologic time period lasts between 23 and 26 million years BCE. Continents collide, mountains form, and the climate cools as major forests turn into grasslands.
Quaternary Period:	This represents the shortest of the geological time frames ranging from 2.3 million years BCE to the present. This period is divided into two epochs:

The Pleistocene Epoch, ranging from 2.6 million to 11,000 years BCE and characterized by at least 50 major climatic oscillations from cold glacial periods that lasted as long as 100,000 years to warmer interglacial intervals that lasted on average about 10,000 years; and

The Holocene epoch, which we are currently in. This period is defined as an interglacial period beginning, 11,600 years BCE and extending to the present. It is in this period when major human civilization appeared along with the modern fauna and flora which we continue to see today.

Note that over the course of the past 400 million years, there have been at least two massive extinction events, both caused by major natural physical events. In the later chapters of this book, we caution about the possibility of such a man-made extinction event that might occur unless we "take up arms" against global warming and climate change.

Now back to our tree's story:

Roughly four billion years ago (i.e. the Precambrian Era), when the only life forms were simple anerobic (meaning living without oxygen) microbes, before there was any evidence of photosynthesis (remember: **CO_2 + H_2 + Solar Energy produces Sugar/Cellulose + O_2**), there was virtually ZERO oxygen in the Earth's atmosphere. Life as we know it simply could not have existed. But there is evidence that as long ago as 3.5 billion years, there evolved simple bacteria called cyanobacteria that were capable of carrying out the photosynthetic reaction using a substance that is similar to chlorophyll, the green substance in plants today that is largely responsible for plant growth. These single-cell bacteria were likely the first precursors of plants and captured solar energy from the sun and absorbed carbon when they took up CO_2 from the atmosphere. The oxygen subsequently released into the atmosphere was really just a waste product of the microbe's energy utilization. But as it happens, that waste product was critical to the development of all the aerobic (needing oxygen) species on Earth.

Starting as long ago as 2.5 billion years, single plant cells, including algae, became much more common and continued to release oxygen utilizing photosynthesis into both the atmosphere and into the sea. During that time, most of that oxygen was taken up by iron, the result of which are the large deposits of iron oxide frequently found on the ocean floor. Very little of the oxygen was released into the atmosphere. At the time, the carbon dioxide in the atmosphere was approximately 4,000 ppm (parts per million), about 10 times what it is today; and yes, the Earth was a lot hotter than it is today.

Over time, a period of hundreds of millions of years, there evolved more and more single-celled, then multi-celled, and even more complex plant life, that led to ever-increasing amounts of carbon dioxide being taken up from the atmosphere and more oxygen being released into it. It took about one billion years (to 2.5 billion years BCE) before there was any significant amount of oxygen in the atmosphere allowing for the subsequent development of more complex plants and animals, including insects, birds, and eventually mammals.

The critical nature of photosynthesis is best summed up by a statement from Robert Blankenship: **"Photosynthesis is the only significant solar energy storage process on Earth and is the source of all of our food and most of our energy resources…"**

In principle, it is so simple and so elegant. Photosyntehsis uses sunshine, water, and carbon dioxide to store energy in the form of carbohydrates (sugar, cellulose, wood) while giving off oxygen as a by-product. If only society could figure out a way to capture that energy directly…

And yes, this challenge will be addressed in later chapters.

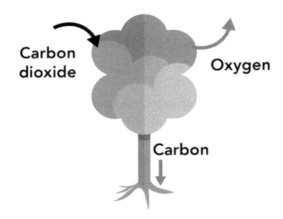

Carbon Dioxide and Oxygen
CARBON CYCLE

FIG 7 THE LIVING, BREATHING TREE

With enormous amounts of photosynthesizing plants sprouting up, two things immediately follow: Atmospheric oxygen increases; and atmospheric carbon dioxide decreases. It has been suggested that dramatic change in both of these contributed to the large evolutionary diversification that has been observed in the Devonian Period starting around 350 million years ago. Some have even taken to describing it as "the first real explosion of life," with the increased oxygen allowing for the further development of aerobic species (e.g., species that require oxygen to grow) and the decreased carbon dioxide resulting in cooler temperatures and a more hospitable atmosphere for a variety of animals to live, prosper, and evolve.

By the time the Devonian Period came to a close, the CO_2 levels in the atmosphere had decreased from over 4,000 ppm before the advent of green plants to under 300 ppm. Note: That's lower than it is today. Simultaneously, over the same period of time, O_2 levels went from virtually zero to an average of 25 percent of the atmosphere. Ever since, that is for the last 300 million years, these values have been relatively stable. In all that time, carbon dioxide never rose above 300 ppm except for the recent spike that began around 1950, and by 2022, it had risen to 420 ppm. Oxygen levels have hovered between 20 percent and 30 percent of the atmosphere dependent on climate situation (e.g., the ice ages and vegetation; again, photosynthesis).

The late Devonian Period gives us an interesting insight into the issue of global warming. By the end of that period (360,000,000 BCE), conditions were ripe, and there had been a massive expansion in plant and animal life, resulting in the oxygen levels in the atmosphere increasing sharply and the carbon dioxide levels (i.e., greenhouse gas) dropping sharply. What followed was an ice age that extended over 100 million years with major polar ice caps and year-round ice coverage extending well south into the southern parts of North America, Europe, and Asia. It's been hypothesized that the ice age occurred as a result of a change in the Earth's orbit around the sun and the sharp drop in atmospheric carbon dioxide causing the Earth to cool. This wiped out much, but not all, of the Earth's animal and plant life

Now this is not intended as a book about how the Earth evolved, so let's fast-forward to the period closer to modern day: the past 100 years.

AND THEN CAME DEFORESTATION

Everywhere I go (I as in the tree), I'm told that I am to blame for a major part of today's world problems. The use of my fossilized ancestors as the major form of energy has polluted our atmosphere and increased atmospheric carbon dioxide, hence greenhouse gases, resulting in global warming. And to make matters worse, my current rate of disappearance as a result of deforestation means that I am less able to remove carbon dioxide from the atmosphere and lower greenhouse gases in the atmosphere. (So in theory I could have been the solution to the problem I created, but not if my family and I are continuously being cut down.) In addition, global warming and climate change have given rise to severe droughts and forest fires around the world. It has also resulted in a huge increase in extreme weather (as in the number of Category 5 hurricanes and other damaging weather events), the loss of marine life, and the extinction of hundreds of thousands or even tens of millions of plant and animal species.

But wait a minute! I can't really shoulder all the blame. Some of the blame rests on the boy for the excessive burning of fossil fuels. And he's the one who has been cutting down trees for centuries. While that might appear a bit simplistic, the fact is that, for the most part, it's true, and we simply didn't give it much thought. Today, we must realize that we are simply addicted to burning fossil fuels and cutting down trees. We didn't feel the impact until after the onset of the Industrial Revolution. Things started to decline in the post-World War II period when we witnessed population growth and industrialization – not to mention the really quite obscene or maybe even catastrophic growth that has occurred over the last 25 years.

FIG 8 THE DEVASTATION of DEFORESTATION

Well, all that's true, but it's also, as you might expect, not that simple. We still have to prove, and we will in subsequent chapters, that the two of us (the boy and the tree) are both to blame for this "hot mess" that is global warming and climate change. There were, in fact, many periods of deforestation and reforestation since I first appeared.

Let's use for example the ice ages that have occurred periodically. There have been many. Wikipedia gives a succinct definition: **"An ice age is a long period of reduction in the temperature of the Earth's surface and atmosphere, resulting in the presence or expansion of continental and polar ice sheets and alpine glaciers."**

Climate change deniers would have us believe that we are simply going through a natural period where the "ice age" is retreating and so the Earth is getting warmer. **This may be true, but it is anything but natural.**

The obvious evidence that the current climate change/global warming we are experiencing is **not** the result of the natural inter ice age (i.e., warming) period is threefold.

1) The normal retreat of an ice age resulting in global warming occurs over a period of 10,000 to 20,000 years and more. Our glaciers are disappearing over a period of 10-20 years. That's about 1,000 times faster.

2) The disappearance of ice ages has never (in 350 million years) resulted in the atmospheric CO_2 going above 300 ppm. Ours is now sitting at 417 ppm (2020 figure).

3) The evidence that global warming is a direct result of human activity is overwhelming with a direct correlation with our use of fossil fuels, our ever-expanding industrialization, and agriculture (deforestation), all of which results primarily from population growth. This all leads to dramatic increases in the burning of fossil fuels, increased release of CO_2, and increased greenhouse gas in the atmosphere leading directly to global warming. The evidence for these positive correlations is overwhelming and will be demonstrated in Chapter 15.

Let's get back to the ice ages. All the many ice ages that have occurred since the Devonian Period (360 million years BCE) and including up to as recently as 10,000 years ago had major impacts on me, the tree. During each glacial period, tropical areas on Earth where I tended to prosper became colder and drier. For example, tropical rainforests changed dramatically into dry seasonal forests and even broad, wide-open savannas. These are mixed woodland and grassy areas where the trees are spaced wide enough apart so that a canopy doesn't form and sufficient light reaches the ground to allow the growth of many different herbaceous species, primarily grasses. There were of course other areas further north and sometimes further south where the ice and cold simply killed my ancestors. In other areas, dependent on the topography and the extent of the climate and temperature change, some or even many of my ancestors escaped the impact of the ice. These areas, though much colder than in the time before the ice set in, offered refuge to many different species trees, plants, insects, and animals. And then in subsequent interglacial periods when the ice receded and humid conditions returned to the tropics, the forests which had escaped the ice were able to expand (the Amazon rainforest is a good example), and they were populated by the plants and animals from the species-rich "refuges." Through each of these periods, untold amounts of trees and vegetation died and were buried only to reappear later on as coal, oil, and gas as man's discovery of, and need for energy sources were expanded.

But the current period of warming is clearly manmade. It started early with major population growth, the establishment of the first large cities, and the clearing of land for construction and for agriculture. And then as the population grew, seemingly unabated, more land was required for agriculture and for construction of cities, highways, and industries, and trees in massive quantities were clear cut without any consideration for the environment; simply to accommodate the population growth of humankind. Clearly this sort of deforestation is different and certainly not caused by any changes in the climate as occurred for example during the last ice ages. If anything, it is one of a number of key causes of climate change. More on that to follow. The second important point is that deforestation continues to occur and with increasing rapidity. To put it very bluntly, it is simply man chopping down trees almost indiscriminately for his or her benefit(s).

FIG 9 THE MULTIPLE SOURCES OF DEFORESTATION

The more common use of the word "deforestation" in modern times asserts that it is the active removal (usually by chopping down, but sometimes by controlled fires) of trees by man. At first, the intention is to provide shelter, food, warmth, and fuel, later it's to create more space for agriculture, for the development of cities, and the ever-increasing urbanization.

While it might be tempting to suggest that most "organized" deforestation has occurred since the start of the Industrial Revolution, that would not be true.

If you were to visualize the ancient Middle East sometime around 3000 BCE, you would see massive forests everywhere largely made up of cedar trees (remember the cedars of Babylon). But by the year 2000 BCE, much of the forests in those areas of the Middle East had been denuded or deforested, with the wood being used to build the majestic palaces and temples of the Mesopotamians and for wood for the mighty ships of the Phoenicians who plied the trade routes around the Mediterranean and beyond.

Another example of major deforestation is the incredible story of Easter Island in the South Pacific. In 800 CE, a group of Polynesian explorers arrived in their outrigger canoes having crossed vast stretches of the Pacific Ocean. They arrived to a lush island called Rapa Nui that was rich in vegetation with enormous palm trees. Over time, they sculpted nearly 900 magnificent monumental structures chiseled from volcanic stone. The monuments averaged 13 feet in height and weighed in excess of 14 tons each and are thought to have been built in honour of the islanders' ancestors. Over a period of time of hundreds of years, the island was denuded of forests. Initially, it was thought that the deforestation must have been caused by "natural" events, but more recently research points to humans as the culprits as poor shepherds of the trees and land.

Jared Diamond in his book *Collapse* describes what happened on Easter Island as an example of environmental degradation and refers to it as "ecocide." It's not the first and certainly will not be the last time that a civilization and a physical geography has been denuded and destroyed by unchecked human activity. Even more certain, humans continue to expand their massive footprints, it's an issue that must be addressed.

TREES...BY THE NUMBERS

The numbers are astounding, and really more than a bit hard to fathom. Estimates are that in 2015, there were three trillion trees worldwide. In 2017, 15 billion trees were chopped down to provide us with toilet paper, timber, and other products. Trees were also chopped down to allow for farmland expansion for agriculture. Trees were clear cut to make way for new housing, new factories, shopping malls, and new roads. By the way, that's about 30,000 trees cut down per minute. Now if you think about it, that's "only" 0.5 percent of the Earth's trees cut down per year. Put in those terms, it doesn't sound particularly significant, but later on, in a section on reforestation, we'll examine just how significant that number really is.

Let's use a different metric just for emphasis. In 2017, some 39 million acres of trees were cut down around the world. That amounts to an astonishing 40 football fields cut down each and every minute of the day. If you're checking some of the math, the number of trees per acre is estimated to be on average 400, although obviously that depends on the species of tree, their age, and how the forest is being "managed."

Today, the greatest and some of the most important loss of forests on Earth is occurring in Brazil. In the month of May (2019) alone, 750 square kilometers of forests were destroyed. That's 75 million trees in a single month, or just under 1 billion trees in a year. And in spite of worldwide concern and even outrage for the destruction of Brazil's rainforests, the numbers have increased by as much as 40 percent since 2016. Using another metric, since 1975, Brazil has lost 760,000 square kilometers of forest. That's almost exactly the same as the total land area of either the province of Ontario in Canada or the state of Texas in the United States.

As you might expect, not all trees or forests are equal, and the importance of Brazil's rainforests for the health of the planet cannot be overstated. Brazil is so important because it is home to 33 percent of the world's rainforests, which play a critical role in the Earth's oxygen and carbon dioxide cycles. It produces 6 percent of the world's oxygen and is thought to play a very major role as a carbon sink, meaning that the forests readily remove large amounts of carbon dioxide from the atmosphere. Yes, that's for photosynthesis and plant/tree growth. The forests, in fact, function as a sink to remove large amounts of CO_2 from the atmosphere, proportionately lowering greenhouse gases and slowing the rate of global warming even in the face of our continued, burning of fossil fuels. **Deforestation lowers the amount of carbon dioxide removed from the atmosphere, thereby adding to the greenhouse gas accumulation in the atmosphere brought on by our continued "love affair" with oil and gas. Pause for just a moment and think how important the Amazon rainforest must be for our very existence.**

This is a good place to remind ourselves of the very global or international nature of global warming, which obviously knows no borders. What happens in the Amazon rainforest in South America has enormous implications for people worldwide, not just those in Brazil. This is the antithesis of the old adage, "What happens in Vegas stays in Vegas." Clearly global warming and climate change represent the ultimate of real global and indeed international issues. There is no escape.

What about tree planting programs? There are exciting new tree planting programs in China and India to go along with programs in North America. The numbers, however, are daunting, and if reforestation is to be a solution to at least part of the problem in global warming and climate change, then the planet is going to have to take these efforts to another level. The numbers, as they stand simply don't add up. You can't just plant hundreds of millions of trees or even billions. You must at the same time put a stop to the massive deforestation that is occurring in South America and elsewhere around the world.

In 2017, while 15 billion trees were cut down, only five billion were planted. That's a net loss of 10 billion trees each year. And remember, it'll take decades before those little seedlings are large and mature enough to do the work of carbon capture of the full-grown tree that you cut down. It should also concern you that as our population grows and more people move into cities and we need more land for highways and factories and of course agriculture, the incentive to cut down even more trees will only increase.

THE CURRENT AMAZON DEBACLE

FIG 10 THE TRAGEDY OF THE AMAZON

Think about it for a moment. On the one hand, the deforestation of the Amazon rainforest is a deplorable travesty, a blight on mankind. On the other hand, it is as the mindless expression goes: "business as usual." The living trees and forests on Earth can be seen as our protectors in the fight to control greenhouse gases and global warming. The forces of population growth, industrialization, and the necessity to feed the world, on the other hand, are driving us to cut down those trees, and at an alarmingly increasing rate. The Amazon rainforest is, in a sense, a bellwether, or, if you will, the canary in the coal mine of how humans and nations are failing miserably in any attempt to address the issues of global warming and climate change.

The Amazon Basin of South America (it's not just in Brazil) encompasses an area of seven million square kilometers, of which 5.5 million square kilometers is rainforest. That's an area that is 2,000 kilometers by 2,800 kilometers. That's more than half of the entire land mass of the United States and almost half of the land mass of Canada, including its most northern area. And it's two to three times the entire land mass of all of the countries of the European Union. So don't picture your local beautiful forest when you're thinking about this problem. The Amazon rainforest is home to more than 400 billion trees, between 15 and 20 percent of the trees worldwide.

In 2018, it's been estimated that 3,500 square miles of Amazon rainforest was cut down. That's 1.2 billion trees and an area approximately equal to 987,000 soccer fields in a single year. And yes, that's very hard to imagine.

There is a danger of looking at the numbers and concluding that while that's an awful lot of trees being lost through deforesting the Amazon, it's still just a drop in the bucket somewhere under 1 percent per year. The concern is that, in a large ecosystem such as the Amazon rainforest, the approach of a tipping point where changes are irreversible may be accelerated. For example, the rainforest, by recycling moisture through its trees and vegetation, is responsible for about half of its own rain and clearly has an impact on both regional and global water cycles. The Amazon rainforest has its own microclimate, which can impact pressure and weather systems worldwide but especially throughout the north Atlantic. So you see that even moderate changes (e.g., the annual rate of deforestation) to the Amazon rainforest can have a major impact on the world's climate. Think about this potential headline: "Major Deforestation in the Amazon Causes Massive Fires in California." Who was it who coined the phrase that we live in a global village?

AND WHY ARE THERE SO MANY TREES BEING CUT DOWN

On the one hand, it's pretty straight forward. There is no fundamental protection; no international treaty or agreement to limit what has really been continuous deforestation on a massive scale. The population base is continuing to grow at a rapid pace, and quite simply, they want stuff, and that stuff, at least in today's world, too often requires that we cut down trees. Virtually no global thought goes into the process.

1) We want wood, and often exotic woods to build furniture and homes, and so we cut down trees.
2) We want meat. The cattlemen need more land to allow their animals to graze and feed, and so we cut down trees.
3) We want more fruit and vegetables that are readily grown in tropical climates, and so we cut down trees.
4) We want more vegetable oils (e.g. palm oil), so we cut down trees in the rainforests.
5) We want to dig to find oil, aluminum, copper, gold, and diamonds, and so we cut down trees to access the ground below.
6) We want to live in big, expansive cities with excellent highways and roadways, and so we cut down trees to clear land.
7) We need water, and we need damns to control the flow of water, so we build damns, and we cut down trees (or drown them).
8) Some of us still use wood to heat our homes or luxuriate in front of our fireplaces, and so we cut down more trees.

The list goes on, and we are doing that at an increasing rate year after year as the demands for all this stuff increase. Corporations and governments use the most expedient way to achieve those goals, and that is to cut down trees. And for economic (it's cheaper), social (the populations are poor), and political (the politicians have short term horizons) reasons, the great rainforests of South America are the target for this continued deforestation and thoughtless exploitation.

At the time when this book was being written, large segments of the Amazon rainforest in Brazil were on fire. Some said that, in fact, they were being torched. The area had become a political battleground between the left and the right. Between the tree hugging conservationists and the climate change deniers, there was a ton of rhetoric, lots of name calling, and little else. Incredible as it might seem, right wing governments around the world scoff at the idea of global warming, and farmers feel free to burn forests to clear them to make way for new soy farms and cattle ranches. Some in this weird world say that forests were being burned down just to make the political point that "it doesn't matter." The worldwide condemnation from many different groups concerned with the environment and global warming attempt to put pressure on the leadership of Brazil. The president of Brazil responds in defiance by admiring the burning forests and assuring the population that clearing the forests is critical for their economic benefit, for the creation of more jobs and more wealth. It's all like a bizarre reality television show.

So, for a quick refresher, what happens when we cut down too many trees?

1) Large scale tree cutting leading to deforestation can transform forested areas to new areas with little vegetation. Trees and green plants take up carbon dioxide and water vapor from the atmosphere and give off oxygen. The increased carbon dioxide in the atmosphere, acting as a greenhouse gas, can result in global warming.
2) Cutting down trees can result in the loss of habitat to tens of thousands (and more) of animal species, causing harm to the ecosystem as a result of the loss of biodiversity.
3) Cutting down trees can lead to desertification, where previous fertile grounds lose the ability to retain moisture in the ground, leading first to drylands then to deserts. Stewardship of the land is always important. Remember for a moment the terrible dust storms in the American West, a period called the "Dirty Thirties." That was at least partly a consequence of deforestation.

4) Tropical rainforests, like those in the Amazon, promote continuous cycles of evaporation and rainfall. Loss of the rainforests will result in warmer and drier climates.

And here again, there's a positive feedback loop, one of many that are detrimental to climate change and our health, and you can start at any point in the loop.

Forest fires result in **increased release of CO_2 into the atmosphere,** which results in **deforestation,** which results in **decreased CO_2 being absorbed by fewer trees in the forests,** which results in **increased atmospheric CO_2,** which results in **increased global atmospheric temperatures,** which leads to **increased evaporation of moisture from the land,** which leads to **decreased moisture in the land,** which, completing the cycle, leads to **increased forest fires.**

* Burning forests, like burning fossil fuels, increases atmospheric CO_2.
** Forests are cleared for agriculture, housing, industries, highways, etc.

FIG 11 THE FOREST FIRE CYCLE

The cycle of forest fires has many different origins coming from lightning to human error/stupidity, from throwing a lit cigarette out the car window to setting fires intentionally to clear land or, increasingly in recent years, to create havoc.

The best known examples in the year 2020 would be the massive brush fires in large parts of Australia, the devastating forest fires in very residential areas of California, and the purposeful clearing of land in Brazil through the active setting of parts of the Amazon rainforest ablaze.

But then let's go back and remember what the tree has done for us. It goes back to Shel Silverstein's book, *The Giving Tree*. Without the tree, it's hard to imagine how we would have developed. What would we look like today without it? First the tree gave us food from its branches, then it gave us fuel to heat our food and provide warmth in winter, then wood to build our homes and our cities. It then gave its remains/fossils, which provided us with energy to drive our cars and develop our industries. And clearly, it's done all that much too well, too easily, and too inexpensively, and for that, we're paying dearly in terms of our fight to save ourselves in the face of serious global warming.

Chapter 7

METHANE'S ROLE

At first blush, a discussion about methane might appear to be a diversion from our story. For many years, we've largely focused on the role of carbon dioxide as the major product of the combustion of fossil fuels and the major culprit in global warming. It is true that the two most prominent of the greenhouse gases, those that we talk about the most, are water vapor (H_2O) and carbon dioxide (CO_2). The water vapor in the atmosphere increases as water is evaporated from the usual water sources: oceans, lakes, and rivers. Evaporation increases especially as the temperature in the atmosphere increases. Note again the positive feedback loop associated with water vapor and a warming atmosphere. Water vapor, functioning as a greenhouse gas, traps the sun's heat in the atmosphere, and the warming increases the amount of evaporation of water from any number of sources, especially open water, causing more water vapor in the atmosphere, more greenhouse gas, and more warming. We'll talk about this in greater detail in a subsequent chapter specifically dealing with climate change. Carbon dioxide is produced as carbon is released from burning wood, coal, or oil, and the released carbon combines with oxygen in the air to form CO_2. Both water vapor and carbon dioxide are especially adept at absorbing the thermal (infrared) energy radiated from the Earth (originating from the sun), thereby increasing the temperature in the atmosphere.

While CO_2 is most commonly thought of as the "bad boy" in the greenhouse gas global warming equation, it may be that methane, the third (following water vapor and CO_2) most abundant greenhouse gas, may become the most dangerous. For a start, methane, while it might appear to be a minor player in terms of the absolute amounts in the atmosphere, can quickly become a major player in that it is at least 30 times more proficient than carbon dioxide at trapping heat in the atmosphere.

Then what is methane, and where does it come from, and why should we be so worried about it? It's a very simple compound: a single carbon atom surrounded by four hydrogen atoms, or CH_4. The primary source of methane is the bacterial decomposition of organic (carbon-containing) material in the absence of oxygen. So, if the truth be told, as I revert to my role as a tree, I'm also responsible for a lot of the methane released into the atmosphere since it comes largely from the decay of long-ago buried vegetation, including trees.

Methane, like carbon dioxide, is emitted into the air by both natural and man-made sources. Methane is an important and also abundant product of digestion, be it from humans or from farm livestock and other animals. One of the largest sources of methane release is from the simple decay of organic waste on farms or municipal solid waste landfills. Plenty of methane is also released into the atmosphere as a result of livestock and other animals, including humans doing "you know what." Methane gas is also emitted into the atmosphere during the production and transport of coal, oil, and gas, even before they are combusted, in order to release their stored energy. In fact, natural gas is 90 percent methane. When it is burned, it also produces carbon dioxide, which is released into the atmosphere along with substantial amounts of methane. Both therefore accumulate in the atmosphere as critical greenhouse gases; sort of a "double whammy" in terms of global warming.

The really frightening and somewhat unknown part of global warming is the role of the melting permafrost in the Arctic and Antarctic regions. Methane, a product of degraded plant and animal material is thought to be trapped in large quantities in the frozen permafrost. The amount of that methane which will be released into the atmosphere as a result of global warming and the melting of the polar ice caps is a very major and concerning question. In a sense, this forms the possibility of another somewhat terrifying positive feedback loop. As global warming allows more methane to be released into the atmosphere, resulting in even more greenhouse gas in the atmosphere, additional warming of the atmosphere, additional melting of the polar ice caps, and subsequent uncovering and exposure of even more permafrost. Talk about a perfect storm that might just be it. Certainly, more attention must be paid to sources of methane release into the atmosphere and of course ways to stop—or at least limit—that release. There is already some evidence that temperatures at the north and south poles are increasing faster than other parts of the Earth and are wreaking havoc with weather patterns around the world. Remember the polar vortices that we've experienced these past winters over large parts of the United States?

FIG 12 MELTING ICECAPS and
THE APPEARANCE OF PERMAFROST

THE FACTS, PLEASE

Throughout this book, we are trying to provide you with facts; with knowledge so that you might have a better understanding of the real perils of global warming and climate change and the crisis facing humanity. What follows is a brief synopsis, or, if you will, a **Coles Notes Version** of a global warming fact checker. (If you don't know what Coles Notes are, google it. It's what many of us used pre-Google). In the appendix, you will find a more complete discussion of "How to Deal with a Climate Change Denier."

THE TRUTH ABOUT TREES AND GLOBAL WARMING

1) Is it true that the Earth has undergone many versions or cycles of warming and cooling?

    Yes. That is correct, but note that all of those occurred as a result of physical changes to the Earth's relationship with the sun or major geological events, such as tectonic movement of large parts of the Earth, even continents, and/or very major, even catastrophic, volcanic events. Today's global warming is occurring rapidly and without question as a direct result of human activity.

2) Is it true that not only has warming occurred in the past, but that it was also generated by increases in carbon dioxide in the atmosphere?

    Yes, that is correct. There have been periods more than 300 million years ago when the carbon dioxide levels in the atmosphere were so low and the temperature of the atmosphere was so low that the variety of lifeforms on Earth were very limited. There were also periods of time when carbon dioxide levels in the atmosphere were so high as 2,000 ppm (parts per million), resulting in temperatures that were so high, life on Earth was limited to a few microbial species. But that was at least one billion years ago. Carbon dioxide levels in the atmosphere have never been above 300 ppm over the past 3,000,000 years.

That is until the past 50-60 years when they had risen to 417 ppm by the year 2020 and are continuing to increase at an alarming rate. That is a problem, and unless it is addressed, it may become an existential question; as in a problem for our very existence.

Those are the two arguments that climate change deniers use to deny that global warming is a result of man's activities. They sometimes concede the fact of global warming, albeit reluctantly, when they are presented with and forced to see just how clear the data is, but they never admit that it is caused by human activity.

3) It is also true that there is a very excellent and direct correlation between global warming, levels of carbon dioxide in the atmosphere, population growth, industrialization, and the burning of fossil fuels.
4) It is also true that over the same time period, our oceans have warmed, and they have become more acidic as a result of higher levels of CO_2 dissolved in the ocean waters.
5) It is also true that while small changes might have been apparent after the start of the Industrial Revolution in the mid-1800s, they have accelerated in the post-war industrialization in the 1950s, and towards the end of the first quarter of the twenty-first century, they have become critical.

Chapter 8

THE RELATIONSHIP SOURS

It's a little hard to say when our relationship, that is between the boy and the tree, really started to sour. You could of course go all the way back to the earliest man, well before the advent of Modern Man about a million years ago. That was when the first primitive man discovered that if he could create fire and burn wood, he could use that released energy to warm himself in his cave or cook his food and make it more palatable. It has not been established whether that first fire was spontaneous and that man simply merely took advantage of a fire that had started naturally or spontaneously, or whether he had the know how to actually light the fire for his own use. Consider just how early in our development that was. It is estimated that there were only some 10,000 to 20,000 humans living during that period. By our standards, we would certainly have been on the endangered species list. Obviously, no one would have thought that burning such a "found item" that is wood in the first instance might be predictive of bad things to come. There is even evidence of the use of fire to cook meat deep inside caves as long ago as 1.5 to 2 million years ago.

Now, you probably expect to read that the real breakdown in the relationship coincided with the start of the Industrial Revolution in Britain in early part of the nineteenth century. That is in all likelihood the most well-considered period, in light of the burning of excessive amounts of coal as a source of energy, the development of the steam engine, and the prominence of industrialization. What we also know is that even back in the mid nineteenth century, even before global warming was coined, the burning of so much coal started to have an important impact on man's health, specifically the condition of his lungs. The gender reference reflects that it was only men who went down into the mines to dig for coal.

There is an interesting historical note that dates back even earlier to England in the late thirteenth century. The air pollution in London was so bad that King Edward took action in an effort to reduce the air pollution produced by the burning of sea-coal. This was coal found on the beaches in England, which had broken off from veins of coal located on the bluffs. He threatened the citizens of London with harsh penalties if they didn't stop burning the coal. Apparently, it was so bad that even his mother had been forced to flee the city to avoid the polluted air. The "Environmentalist King," as he is referred to today, is reported to have said to his mayors and sheriffs: "The air there is so polluted over a wide area...to the detriment of the citizen's bodily health." He banned the burning of coal and ordered the use of alternative material, such as wood. Nobody listened. He would certainly fit in with today's officials who are concerned with environmental issues around the world, but are rarely listened to, and as a result have been largely ineffectual in producing change.

Many changes started to occur around the time of the first and second (1800-1850) Industrial Revolutions, which exacerbated the coal problem. Demographics changed dramatically during this period as the population grew and people migrated cities, and obviously major changes occurred to the relationship between the boy and the tree.

Year	Population of Great Britain	Population of London	(%)**
1250	4,200,000	18,000	0.5%

1350	2,600,000	80,000	3.0%
1450	1,900,000*	50,000*	2.6%
1550	3,100,000	90,000	2.9%
1650	5,300,000	400,000	7.5%
1750	6,500,000	760,000	11.7%
1850	15,300,000	2,200,000	14.4%
1900	32,600,000	6,600,000	20.9%

- The sharp drop in the population between 1350 and 1450 is attributed to the Black Plague, which reached England in 1348 and returned repeatedly through 1400, killing about a third of the population, more than one million people over that period.
- The dramatic increase in the percentage of the population of Great Britain living in London demonstrates the process of urbanization, which was to have huge impacts worldwide on how we live, how we work, and how we procure our food, presenting a whole host of other changes and difficulties.

The story of the breakdown of our relationship, if the truth be told, has little to do with the tree but rather with the boy's proclivity to reproduce and continuously desire more. Just look at the numbers above. Not only did population numbers grow ridiculously around the time of the Industrial Revolution, but so did the use of coal to drive not only population growth, but also for society's penchant to produce new industries.

Coal Production in Great Britain

1600	0.1	million tons
1700	2.5	million tons
1800	50	million tons
1900	250	million tons
1950	260	million tons

1986   100   million tons
2018   6   million tons*

* You might have first seen this sharp drop in the use of coal starting after 1950 and quite spectacularly by 2018 as a bit of great news. Unfortunately, that's hardly the case. This precipitous drop in the production of coal in Britain didn't represent a drop in the use of fossil fuels or a drop in greenhouse gases produced, but merely a dramatic shift from coal to oil and gas. Indeed, the burning of oil and gas continues unabatedly as the energy source to drive the lives of Britons and their industries.

So, you see, by the year 1900 Great Britain was faced with a major problem, and it was only going to get worse… and not just in Great Britain. Great Britain is a clear example because it is generally considered as a major seat of the first major Industrial Revolution. If not in Great Britain, it surely would have occurred elsewhere; and of course, in short order, it did.

The first commercially available oil wells were established in the United States around 1860. While it was slow to replace coal, the oil was useful in allowing the production of new oil-based products like kerosene, which quickly became a popular, cheap, and supposedly clean fuel for lighting homes. As is the case for many major technological advances, the military played an important role in the conversion from coal to oil. Both the British and American Navies were quick to convert their battleships, realizing that oil produced much more energy per pound than coal. In fact, it was the military that was driving force for the move from coal to oil.

In a very real sense, the dye was cast. The boy had committed to population growth and industrialization, and the tree's ancestors would provide the fuel or energy. That didn't bode well for the future, but for the most part, we were blind to that until quite recently.

Chapter 9

THE ANTHROPOCENE

"The Anthropocene is a 'proposed' geological epoch dating from the commencement of significant human impacts on Earth's geology and ecosystems, including, but not limited to anthropogenic climate change."

The derivation of the word Anthropocene, as defined by *Smithsonian Magazine*, comes from the word "anthropo," for "man," and "cene," for "new," the notion being that human activity is now a dominant force on influencing the climate and the environment. The negative connotation is that human activity has resulted in mass extinctions of plant and animal species, polluted the oceans, and altered the atmosphere, among other lasting impacts.

Much has been written about this relatively recent concept. It is a popular idea viewed and adopted by many, especially those with well-developed concerns for the environment. The concept, accepted by some but certainly not all, is that we have entered into a new geological age, an age in which human activity is having a profound impact on shaping the Earth, especially in respect to its physical structure and diversity. At first glance, you might say, "But of course! That's obvious." But before doing that, we should review the concept of geological age or the geological time scales as defined by organizations such as the International Commission on Stratigraphy (ICS) or The International Union of Geological Sciences (IUGS). The geological time scale, or GTS, is a chronological dating system that describes changes in the Earth's geology (strata). Earth scientists have used these strata to describe the events and the relationship of those events to one another, which have occurred throughout the history of the Earth. Each of those periods has been designated as an epoch.

We are currently, according to the geologists who create these definitions, in the Holocene Epoch, which stretches back about 11,650 years, from the retreat of the last major glacial age to the present. The Holocene Period also coincides with the major growth of the human species, at least in numbers and their impact worldwide. Before that, we had the Pleistocene Epoch, which lasted from 2,600,000 years ago to 11,650 years ago. It is also commonly referred to as the ice ages. It is a period where many species of plants and animals became extinct, especially in the northern hemisphere. It is also the same period where many of the large mammals, including all species of dinosaurs, and also Neanderthals, became extinct.

As of June 2019, the powers that be (primarily the geologists at ICS and IUGS) had not yet "sanctioned" the new epoch, the Anthropocene, even as some have been attempting to assign a start date, while others have been using the concept and name quite liberally. The naysayers suggest that we don't have sufficient evidence that we have started into a new geological event in terms of "geological strata" on Earth. To others, it appears pretty clear that the geology of the Earth has undergone a major change.

One "only" has to view the pictures of the Earth from space or even from the birdseye view from an airplane to understand the enormity of man's impact on the Earth. Concrete jungles are now so prevalent, they have replaced natural lands and more recently agricultural lands. When did it start? Was it at the onset of the Agricultural Revolution, when man started to make a major mark on the Earth some 10,000 years ago, or around the beginning of Industrial Revolution in the 1850s when the burning of fossil fuels started to become important? Or was it around 1950, where man's hand in global warming as a result of burning of those fossil fuels and the resulting change to our climate became so significant and irrefutable? Some have even used the term "Homogenocene" to describe this new epoch where man's "footprint" on the physical characteristics of the Earth has clearly become unavoidably obvious. The arguments are more than a bit academic in nature, and as Shakespeare said, "What's in a name?"

There are a number of excellent images that you might want to think about or even imagine when considering the Anthropocene. These are images that hit home in terms of our/man's impact on the home we call Earth. There are now many books and movies that have been made to describe the Anthropocene, some very rational and scientific in nature and more than a few that are apocalyptic. Some merely show changes in the Earth's surface while others depict tragic poverty and filth.

FIG 13 A PICTURE OF THE ANTHROPOCENE

Here are some additional imagery for you to consider.

1) The massive dams, which man built worldwide and which now control the flow of water in many if not most of the earth's major rivers.
2) The polar bear in search of stable ice flows from which to hunt.
3) The view of the Earth from the International Space Station, and the cities lit up at night.
4) Corn fields and wheat fields in the west that stretch as far as the eye can see.
5) Immense mountainsides that have been clear-cut by the lumber industry and huge parts of the Amazon deforested and sometimes viewed partially ablaze.
6) The concrete jungles—buildings and factories, roads and highways in so many cities around the world that have supplanted previously green space.
7) Massive garbage dumps indicative of our excesses and our inability to either consider or deal with the massive amounts of waste that we produce, sometimes needlessly.
8) Huge pockets of air pollution around the world, take specifically the image of Beijing during their last Olympics.
9) The Great Pacific Garbage Patch, and other disgraceful evidence of massive pollution seen in so many parts of our oceans, lakes, and rivers. And why, pray tell, is the color of the water in so many of our waterways so often a filthy brown?
10) The dire poverty and images of dying children in so many parts of the undeveloped world with altogether depressing scenes of famine, drought, and political unrest.

These are all images that we have seen repeatedly in recent years. Note that not all are ugly; some are quite beautiful, but all represent evidence that we have, indeed, entered into a new era of man's incredible impact on the physical nature of our Earth, regardless of whether it is officially deemed a new epoch or not, the Anthropocene.

There has been much written and produced about the Anthropocene. Documentaries by National Geographic and articles such as: **"Age of Man: Enter the Anthropocene"**; and a collection in the resource library of National Geographic with the distressing title: **"Extinction."** They describe in great detail the complete disappearance of distinct species of plants, insects, and other animals from the face of the Earth. The process of extinction has always occurred, but slowly over time. In **"Our Era,"** the extinction process is incurring at a vastly increased rate.

If it is true that the phenomenon that is the Anthropocene has some direct correlation to the idea of extinction, it becomes clear that mankind has some serious work to do in order to correct the path we're on. If there has ever been an understatement, that must be it.

In a sense, it might be viewed as if there are two quite distinct depictions of the Anthropocene: The good representing man's continued development and presence here on Earth; and the bad representing all that is negative (and dirty, to say the least) in terms of mankind's poor and often disgusting shepherding of the environment, including the air, the water, and the land. The first includes the massive changes that man has made to the land, the major excavations made to build cities and roadways, and the massive clearing of land to make way for the agricultural needs of an ever-growing population. They also include the major engineering feats that have been made to "mine" minerals and fuel and to control the flow of waters for both power and agriculture. The second of course involves our too often complete failure to take good control of those major excavations and the massive and seemingly permanent pollution and ugly scars that are a constant by-product. It's hard to fathom how we tolerate the continuous and monstrous build-up of garbage in landfills and often in random, completely open areas in more undeveloped parts of the world. Our apparent lack of respect for our waters as we allow them to be polluted by industry and by organic wastes from agriculture is so apparently self-destructive. And then of course there is our atmosphere as it is repeatedly fouled with noxious sulphur, phosphate, and nitrogen-based chemicals spilling out of fouling industries. And then we get to the less percievable increase in greenhouse gases, such as carbon dioxide, methane, and water vapor. These are a result of our continuous dependence on fossil fuels, which are central to the major theme of the book, and that is global warming and climate change.

Chapter 10

DEFINING GLOBAL WARMING

In this and the chapters that follow, we get to the central theme of the book; the nitty gritty and indeed the **proof** of the direct relationship between burning fossil fuels and the **dangers** of global warming. Some would suggest and have us believe that the relationship is complex and unclear. Well, that simply is not true. They may be complex, but they are very clear. We must be alarmed and on guard to those untruths that are so often foisted upon us by those who have strongly vested interests in the continued dominance of the fossil fuel industries. This cannot be allowed to continue, and here is why. It starts with a series of events, described below. And then we'll add the timing and the actual numbers to complete the picture and to demonstrate without equivocation that the process is serious, even critical, and that it is manmade.

To be fair, one of the problems is that on a day-to-day basis, we don't "feel" the issue of global warming, so it is difficult for the average person not immersed in these issues to fathom just how serious the situation really is. The main reason is that the changes in temperature are usually so subtle, often only a couple of degrees or even less, and not necessarily easy to "feel" year to year or season to season. Nevertheless, the ramifications of those changes are enormous. For example, the dramatic increase in severe weather events that many across the globe have come to experience with increasing frequency and often with devastating effects. Make no mistake about it, global warming is real, and the impacts are getting worse every year.

Here are the cascading series of events that result in manmade global warming.

1) Man discovers fossil fuels.
2) Populations increase.
3) Industrialization occurs.
4) Fossil fuel burning expands.
5) Urbanization increases.
6) Deforestation occurs.
7) Greenhouse gases increase.
8) Global temperatures increase; slowly at first.
9) Ocean temperatures change.
10) Oceans and lakes become more acidic; whole bodies of water "die."
11) Extreme weather events occur more frequently.
12) Agriculture is threatened.
13) Glaciers retreat
14) Arctic ice sheets shrink.
15) Antarctic ice sheets shrink.
16) Sea levels rise.

These are not speculative events. They are not what ifs. They are all real occurrences; real events. The implications of these 16 points are all evident; devastating to some (e.g., extreme weather, as in fires and flooding) and worrisome to others. If nothing else, the impacts of extreme weather events and rising sea levels are being felt by billions across the globe.

Don't forget that data is important, and hard numbers don't lie.

In the following pages, we present the actual data that will allow us to understand why and where we are in terms of global warming.

WORLD POPULATION

Year	1800	1850	1900	1950	2000	2020
Population	1 B	1.2 B	1.6 B	2.6 B	6.1 B	7.7 B
Increase/Year		4.0 M	8.0 M	20 M	70 M	90 M
% Increase/Year		0.4%	0.7%	1.2%	2.6%	1.5%

This is one of the essential points that shows how humans are directly involved in many of the changes our Earth is experiencing, and it begs the question as to how growths in the human population can be sustained. Paul Ehrlich in his book *The Population Bomb*, written in 1968, warned us of the consequences of unfettered population growth amidst finite resources. It appears that we haven't been paying attention. The slow-down in population growth is nothing to get excited about, as we are still likely to reach or even exceed a population of 10 billion by the year 2050. You might ask from Paul Ehrlich's book written more than 50 years ago, how have we made it this far? The initial answer would likely be technology and the many efficiencies that it has afforded us in a sense, making our lives easier. We as humans have learned to adapt in many ways. But now the bill is coming due as many of the impacts (e.g., fires, hurricanes, droughts) get increasingly out of control.

URBANIZATION - % OF POPULATION LIVING IN CITIES

Year	1800	1850	1900	1950	2000	2019
World	7.3		16.4	29.6	47.3	55.8
USA	6.0	15.4	39.9	64.2	79.2	83.1
China	6.0		6.6	11.8	34.9	58.9

The rate at which people have moved from rural to urban settings over the past 200-plus years is quite astonishing. In China, for instance, currently there are 20 million people moving into cities every year. That's a mindboggling number. It means that each and every week, approximately 400,000 people relocate from rural to urban settings. Consider how that changes the equation in terms of deforestation, housing, transportation, employment, agriculture, food distribution, water supply, garbage and sewage disposal, and all the other changes or adaptations that occur as large groups of people move from rural to urban living.

FOSSIL FUEL PRODUCTION

Year	1800	1850	1900	1950	2000	2020
COAL (billions of tons)	0.01	0.12	0.74	2.2	6.8	5.5
OIL (billions of barrels)	0	0	0.5	4.5	26.6	35.2
GAS (billion cubic meters)	0	0	0	500	1,084.0	1,650.0

In the western world, oil and then gas following quickly began to replace coal as our principal energy source just after World War II. Coal remained the major energy source in less developed countries such as China until early into the twenty-first century. For a while, we worried about oil being a finite resource, but over time, we began to realize that the notion that oil would only last us for another 15 to 20 years is false. In fact, we've always been able to discover more oil (and gas) as needed to the point where oil and gas deposits, for all intents and purposes, appear to be unlimited. We are now well aware, or should be, that a seemingly unlimited supply of a relatively inexpensive energy source in the form of a carbon-based fossil fuels is available. But having said that, it's now abundantly clear: **this is not a good thing**.

FOSSIL FUEL EMISSION (Billions of Metric Tons of Carbon/year)

Year	1800	1850	1900	1950	2000	2020
Carbon Emitted	0.05	.320	2.0	5.5	25.4	35.0

This really is the crux of the issue, and it's the result of our almost exclusive dependence on burning fossil fuels (coal, oil, and gas) as our primary sources of energy. First of all, note that a billion metric tons is 1,000,000,000,000 kilograms or 2,200,000,000,000 pounds (as in 2.2 trillion pounds). That's a lot of carbon being released even though not all of it remains in the atmosphere. A significant fraction gets dissolved in the oceans and other seas. That's also bad since when the CO_2 is combined with H_2O, it forms carbonic acid (H_2CO_3), and the waters become very acidic with catastrophic impacts on marine life. A significant amount of the CO_2 that is emitted is taken back up by plants and trees for photosynthesis. That would be a good thing except for the fact that we keep on cutting down more trees, resulting in more of the CO_2 remaining in the atmosphere. Each time a molecule of emitted CO_2 remains in the atmosphere, it acts as a greenhouse gas, capturing a bit of the sun's radiant heat and contributing to global warming. Note that these numbers only reflect carbon emissions and not any of the other greenhouse gases (e.g., methane) or noxious pollutants, such as sulphur dioxide or nitrous oxide amongst others. These of course make the increase in greenhouse gases even more alarming.

CARBON IN THE ATMOSPHERE (ppm)

Year	1800	1850	1900	1950	2000	2020
Carbon Levels	280	285	295	310	360	417

This is, at least for the moment, the major culprit in global warming. *Quite simply put, the last time atmospheric CO_2 was this high, humans didn't exist.* Between 3,000,000 BCE and 1800 CE, carbon levels oscillated between 200 ppm and 280 ppm, coming and going largely as a function of the ice ages; the lower carbon levels being associated to periods of low temperatures, glaciation and extensive ice coverage and the higher levels with periods when the ice ages or glaciers were in retreat. With major growth in the population and industrialization post World War II, the carbon released into the atmosphere increased sharply after 1950.

CHANGES IN AVERAGE GLOBAL TEMPERATURE*

Year	1800	1850	1900	1950	2000	2020	2022
	-.25	-.20	0.0	+0.1	+0.25	+0.78	1.13

*   These temperatures are given in degrees Celsius with an arbitrary date of 1900 taken as the base or sort of ground zero. Multiplying these numbers by 1.8 would allow you to express them in degrees Fahrenheit so that the value for 2020 would be 1.3 degrees F.

These values reflect the changes in average global temperatures expressed in the number of degrees Celsius relative to the temperature in 1900, which is assigned a 0.0. This is the standard used by the United Nations' Framework on Climate Change. The temperature reflects the global temperature measured just above the surface of the Earth and averaged over many thousands of measuring sites around the globe throughout the year, so it is a true annual average. The numbers may seem small, but they are not. The rate of change in the last 25 years is roughly 1,000 times faster than previous periods of global warming, having occurred with the retreat of the ice ages. Furthermore, at the current rate, the temperature is estimated to be +2.0 degrees Celsius warmer (that's 3.6 degrees Fahrenheit) by 2050, which would have many devastating effects on the climate, but the impact on the melting of the polar ice caps and sharply increasing ocean levels would be most pronounced. In 2017 Jeff Goodell wrote a terrific book on the topic: *The Water Will Come: Rising Seas, Sinking Cities and the Remaking of the Civilized World*, including fascinating data and stark warnings.

You will have noticed the trajectory of each of these tables; fairly flat between 1800 and 1950, then showing a substantially steep climb to the current values in 2020. This is sometimes referred to as a hockey stick effect, where the phenomenon that you are measuring increases slowly and then at some point increases quickly almost as if there were a tipping point. In fact, there was such a tipping point, and it was the massive industrialization that took off just a few years after the end of World War II, i.e., 1950. There are other factors that show very similar trajectories, including (we won't bore you with all the added data, but these are important parameters that are either contributing to global warming (the first two) or occurring as a direct result of global warming):

Industrialization.
GDP (gross domestic product).
Expansion of agriculture.
Increases in sea levels.
Dissolved CO_2 and oceans.
Thinning of the ice caps on both the northern and southern poles.
Acidification of the oceans and lakes.
Disappearance of mountain glaciers.
Extinction of thousands of plant, animal, and insect species.

The increase in each of these parameters should be of great concern for our futures. The acidification of our waters and particularly oceans are having dramatic and negative influences on a wide range of marine life. There are numerous fresh water and saltwater fisheries that have almost disappeared over the past two decades, and it's not only due to overfishing. The coral reefs seem to be the proverbial canaries in the mine shaft, and they are being destroyed at a rapidly increasing rate. And the list goes on.

So, there it is. It's just a small slice of all the data, but reviewing it, how can you possibly suggest that global warming is a natural phenomenon and that it is not directly caused by man? If you are an argumentative type, you might call the evidence circumstantial. But then you would still have to admit that the circumstances are all pretty clear and waiting to take corrective action short-sighted in the extreme.

My god (that's probably a sun god we are pointing to in exasperation), but it's absolutely amazing and not a bit depressing that this section has to be featured in this story. There are enough of you out there who are in denial; maybe that's just the negatron in you. The Urban Dictionary defines a "negatron" as one who continuously views all or a group of specific ideas in a continuous and negative manner. Or maybe you have bought into the rhetoric of some of those conspiracy theorists; you know the type, who don't believe in vaccines, and the same ones who think that genetically modified organisms (GMOs) are the product of the devil or some evil industrial power. Or maybe, you just hated your science teacher in sixth grade and have despised science ever since. Maybe you buy into the notion that the Earth is really flat. Spoiler alert: it's not.. But if you're here and you're still reading, then that's a good thing.

The concept of "knowing thine enemy" is an important one. What follows are some suggestions about how it is that some apparently sane people, not necessarily the crazies, choose not to accept the reality of climate change. Always keep in mind the important newly popular cliché originally attributed to American Senator Daniel Patrick Moynihan:

"You are entitled to your own opinions...but not to your own facts"

FIG 14 IF IT LOOKS, QUACKS, AND SMELLS LIKE A DUCK...IT'S A DUCK. THE SAME HOLDS TRUE FOR GLOBAL WARMING & CLIMATE CHANGE

So what gives? Just before we produce some answers to the climate change deniers both hardcore and slightly leaning in, let's ask our leadership. Why is it that so many of our leaders, many in positions of political, social, and corporate power appear to be so silent, so oblivious to the facts, and potentially disastrous ramifications of the current climate change crises? Yes, they attend rallies, and release statements of concern and suggest actions that must be implemented by the year 2050 and maybe earlier. And then they quickly defer to the much softer, much less disruptive, and less expensive actions that might be taken sooner, say by the year 2030. But why is there so much talk and so little action? Here are three essential reasons.

1) Why have major public corporations not seized the opportunities to make a difference on climate change issues? First, and foremost, public companies are concerned with their quarterly reports – their bottom line. What did their profit and loss statement say, how did they do over the course of the last three months, and what's in store—for the next three, months? What has been their return on their most recent investments? These captains of our economy, industrial leaders, and politicians have come together repeatedly over the past 30 years to discuss issues of climate change and global warming. First in 1992, there was the United Nations' Framework Convention on Climate Change (UNFCCC). Then in 1997, nations gathered to sign an international agreement, the Kyoto Protocol, which was meant to "operationalize" the UNFCCC by committing industrialized countries to limit and reduce greenhouse gases emissions in accordance with agreed individual targets. The results were considered mixed at best, and as we know, greenhouse gases continued to accumulate. The nations got together again, and now included developing countries in addition to the industrialized countries, to ratify the Paris Accord. The agreement, signed in 2016, is a commitment made within the UNFCCC dealing with greenhouse gas emissions mitigation, adaptation, and finance. The Paris Accord set to improve upon and replace the Kyoto Protocol. The Paris Accord was signed by 197 nations and

ratified by 187. Among other things, it set specific objectives and timelines to reduce our reliance on carbon-based fuels by the year 2050. Notoriously, in 2017, President Donald Trump withdrew the United States from the Accord. This had a terrible impact given the fact that the United States was expected to take a major leadership position in climate change. In 2021, newly elected President Biden, as one of his first proclamations, re-entered the Accord. But alas, it's still only been lip service . Hopefully real actions will follow.

Tragically, in so many ways, the worldwide COVID-19 pandemic and the Russian invasion of Ukraine seems to have diminished the public urgency to deal with issues of global warming and climate change. That, and the fact that targets of the Accord are generally not being met has been terribly disappointing but also maybe not so surprising. On the one hand, the target date (2050) is less than 30 years off. But on the other hand, it's five or six corporate and/or political lifetimes away. Talk is often cheap and easy under such circumstances, especially when you're not the one who is going to have to make the tough decisions. It's not hard to make commitments when you know that when the shovel hits the dirt, you're going to be long gone.

2) Why has it been such an apparently impossible task to take steps to abide by the Kyoto Protocol and the Paris Accord? It's likely quite simply the issue of strongly vested interests. In many instances, the very same people who are charged with implementing the changes are those individuals from industry and from government who profit most from a strong oil and gas energy sector. One only has to follow the money from the fossil fuel-based industries through the long list of lobbyists, consultants, leaders of banks, and politicians to recognize the vested reasons and conclude this is not going to work unless other pressures are brought to bear. Think for a moment what it means to close the fossil fuel industry and the impact on a whole range of activities from heating our homes, driving our cars, producing our food, and driving 90 percent of all of our industries.

Think about a province like Alberta in Canada or a state like Oklahoma in the United States where well over 40 percent of their economies are directly related to the oil and gas industry. When you consider Alberta and add in their agriculture and forestry industries, it adds up to well over 50 percent. Moving to a net zero, or a negative carbon footprint, would effectively devastate those economies, unless of course strong alternatives were to be established. And yet, as we've established and as we'll show in Chapter 16, these changes must take place because the alternative may be nothing short of extinction. There again, a doomsday scenario that none of us want to hear, but as history has showed, denial does not make problems disappear.

3) Education, especially in the science of global warming and climate change, is hugely inadequate. Many of our business and political leaders, while likely not amongst the climate change deniers whom we often berate, are simply not well versed on the critical areas of global warming and climate change. When confronted by powerful climate change deniers, they crumble and insist that while they understand the problem of climate change, they have to pay more attention to the needs of the economy and of employment. How incredibly short sighted is that? Their ignorance results in a certain fear of science, including the data associated with it, this problem is detrimental to the very nature of our society, which values education and progress. And it is on science and on the data where important decisions, such as those associated with climate change, must be made. In a very real sense, this book is also for them; for them to understand that the science is not very complicated, that the evidence is, in fact, crystal clear, and as we'll see in later chapters, solutions are both real and possible.

And finally, we have to recognize that we are not looking at small initiatives where small behavioural changes might result in solutions. The problems and solutions are large and often global in nature. But it is critical to understand that they are not insurmountable. The issues are will and leadership. How do we develop the collective will to change when the change is essential but not necessarily immediately palpable? And where does the leadership come from when the decisions to be made are difficult and often expensive. Below, you'll find three examples of the types of "will" and "leadership" that have, in fact, been demonstrated: two in response to existential threats, and one as a major national challenge of a technological nature.

Stopping and reversing the destruction of the ozone layer was critical for the health of all mankind. It turned out to be a smallish problem, but it demanded international cooperation. And it was very successful.

Putting a man on the moon was a massive endeavour without a clear line to success because much of the new technology to reach the moon had yet to be developed. But the technology was developed, it was in many ways the birth of the digital revolution, and it forever changed many of the aspects of how we live and work.

The largest and most expensive effort ever put in place were the investments made by the Allies, largely the United States, Great Britain, and Canada, to destroy the Nazis, win World War II, and keep a substantial part of the world free.

THE EARTH'S OZONE LAYER PROBLEM

Ozone, or O_3, is a highly reactive gas composed of three oxygen atoms as opposed to the normal two atoms (O_2), which make up the normal oxygen we require to sustain life and which is so abundant in the air we breathe. Ozone occurs naturally in small amounts in the upper atmosphere called the stratosphere, about 10 kilometers above the surface of the Earth. The so-called ozone layer is critical in that it absorbs most of the sun's ultraviolet radiation. In 1976, a group of atmospheric researchers started to realize that the ozone layer, sometimes called the ozone shield, was being severely depleted by specific chemicals accumulating in the stratosphere that were released by various industries. The offending chemicals, which were producing the thinning or depletion of the ozone layer (sometimes also called holes in the ozone layer), were chemicals called chlorofluorocarbons (CFCs), which includes freon, commonly used as a refrigerant as well as an aerosol propellant. It was determined that the thinning of the ozone layer and the subsequent increase in ultraviolet radiation reaching the Earth's surface would be a serious threat to life on Earth, including and especially resulting in an increase of skin cancer in humans, and fast and immediate action was required.

There was an outpouring of concern from the public and from physicians. In 1978, the United States, Canada, and Norway enacted absolute bans on CFC-containing aerosol sprays. The European Union rejected this approach, and CFCs continued to be used in refrigerants and other industrial uses until the discovery of an actual "hole" in the ozone layer over the Antarctic in 1985. Then an international protocol (the Montreal Protocol) was established, putting in place a halt to the production of all CFCs after 1995. By then, concern about the thinning of the ozone layer was so widespread that the Montreal Protocol was signed by all 197 countries. Not only was the treaty signed, but it was also enacted upon by all industries using CFCs. By 2003, scientists announced that, as a direct result of that treaty, the global depletion of the ozone appeared to be slowing. In 2016, scientists reported that there was even an ongoing trend to "healing" the protective ozone layer, a positive outcome to a serious threat to life on Earth. It was a very major threat, and thankfully, it was effectively addressed on a global scale. But compared to global warming, the fix was relatively easy and quite inexpensive. The vested interests stacked against the worldwide ban on CFCs were insubstantial, and the adjustments were made rapidly. Nevertheless, it was a significant victory indicative of what cooperative decisions and treaties might reasonably be expected.

KENNEDY'S MAN ON THE MOON PROJECT

This is essentially an example of outstanding leadership and how the decisions of a single individual with appropriate support can cause major things to happen. To think of President John F. Kennedy's proposal, especially at the time, to be anything but audacious is an underestimate. He put forward a plan to implement incredible disruptive technology that was going to change lives. But at the time, much of the how and what were still to be determined. On May 25, 1961, President Kennedy stood before the United States Congress and proposed that the United States "should commit itself to achieving the goal, before this decade is out, of landing a man on the moon and returning him safely to the Earth." The mission was indeed achieved as Neil Armstrong landed the Apollo Lunar Module Eagle on the moon's surface on July 20, 1969, declaring famously:

> **"That's one small step for man, one giant leap for mankind."**

By all standards, President Kennedy's decision to place a man on the moon was very expensive and viewed by some as quite preposterous. The Apollo's program total cost was $25.4 billion, which in today's dollars would be in excess of $150 billion or as much as $600 billion if expressed as a share of current GDP. Someone writing for MoneyWise, cynically pointing to the "Astronomical Costs of the Apollo II Moon Landing" once said: **"It was one small step for man, one giant bill for America."**

But in retrospect, the naysayers to the United States' space efforts were monumentally wrong. Much has been written about why President Kennedy decided to put a man on the moon, about his desire to improve the psyche of the American people and allow them to "best" their cold war rival, the Soviet Union. The Americans were thought to have been trailing the Soviet Union in the race to dominate space, and they had recently suffered humiliation in the failed Bay of Pigs invasion of Cuba. But President Kennedy spoke about science and exploration; about new knowledge to be gained and new technologies to be developed. He stated:

> **"We choose to go to the moon in this decade and do the other things, not because they are easy, but because they are hard; because that goal will serve to organize and measure the best of our energies and skills…"**

And how right he was. Not only did the Americans win the race to put a man on the moon, but the effort itself spawned untold numbers of discoveries, products, and industries. When President Kennedy announced the mission to the moon, most of the technologies needed to get to the lunar surface and return didn't yet exist. Within those eight years, virtually the entire digital revolution was birthed. These were the very technologies developed that were used to power the spaceship and to guide it to the moon and back.

FIG 15 MAN LANDS ON THE MOON
(Man's capacity to develop new technologies is enormous.)

WINNING WORLD WAR II
Winning the Second World War of course had a different imperative than going to the moon. It was a fight for freedom, a fight against the tyranny and domination of Nazi Germany. In a sense, just like the war we must wage on climate change, the commitment to win World War II represented an existential battle. There would be virtually nothing that the Allies wouldn't do to win the war.

For the purpose of making an argument, let's look at the cost of the war solely from the point of view of the United States. It was of course a larger effort, especially on the part of Great Britain where the losses to Germany in people and dollars were both colossal. The war lasted fewer than four years, but it was not only the most expensive war, it was also the most expensive undertaking in the history of the United States. In today's dollars, the war cost well over four trillion dollars. In the last year of the war, 1945, the budget for defense amounted to about 40 percent of the gross domestic product or GDP of the United States. As an example, in 2017, the US defense budget was just under four percent of its GDP. The worldwide threat from Nazi Germany was viewed by many as existential, and the decision to go to war and make that major commitment was not all that difficult. The sacrifice was substantial on the part of the average American. Meat, milk, and clothing were tightly rationed. Families were limited to three gallons (11 litres) of gasoline per week, and the manufacture of consumer products, such as vacuum cleaners and virtually all kitchen appliances and other home appliances, were banned until the end of the war. As it turned out, not only did America and the Allies win the war, but they got an important return on the investment, as the economy of the United States was exceptionally strong for many years following the war.

The war we waged in 2020 and through most of 2021 into 2022 against the **COVID-19 pandemic** also highlighted that when there is a will, there is a way. Western governments, the United States, Canada, Great Britain, and the European Union invested in excess of $10 trillion to fight the virus and keep their societies intact. These numbers are not far off from what would be required to fight a war against climate change. More about that in subsequent chapters.

WINNING THE WAR AGAINST CLIMATE CHANGE

There are many who believe, ourselves included, that our very existence is seriously threatened unless we can win the war against climate change. Put another way, unless we can curb the tide of global warming by decreasing the release of greenhouse gases and even reverse some of the damage that has already occurred (e.g. the partial melting of the polar ice caps), many life forms, man included, could very well face extinction. If such were even remotely the case, then we should certainly be gathering all the necessary resources to embark an all-out war to reverse global warming. On the one hand, it's very encouraging to know that we have all of the weapons necessary at our disposal to fight and win that war, just like we were able to gather the will, the strength, and the resources to win World War II. But on the other hand, it's terribly discouraging to realize that, for the most part, our industrial and political leaders either seem oblivious to the problem or are so wrapped up with the vested interests of the fossil fuel industries that they are frozen (bad pun) and unable to act.

In a sense, many individuals, institutions, and even countries around the world have been preparing for the battle. There has been lots of talk, which amount to baby steps in the right direction, curbing plastic bags, recycling, encouraging everything green, a beginning to the re- birth of the electric car, and other often symbolic moves. Many smallish investments have been made in alternative energy sources, such as wind and solar, and establishing an international effort against climate change represented by the Paris Accord and other multinational agreements. These are all important, but they are all only drops in a bucket and only set the stage for what has to come. The real effort will be hugely expensive, resembling the efforts that the United States and Great Britain invested to win World War II. It will require an enormous national and international will, but it can be done. Unfortunately, the realization that we should be going to war earlier rather than later against the ravages of global warming and climate change is not yet there. For the moment, the political, social, and economic will to wage, and win that war are not in place, even if the costs and implications of putting off the inevitable make the task more difficult and more costly. But the world will come to that realization, and we should hope and maybe even pray that there will still be time to win it.

Chapter 11
FACT CHECKING THE CLIMATE CHANGE DENIERS

This is a good point to remind ourselves that we are, in fact, waging two wars at the same time. The first is the war against climate change. The second, which is turning out to be more complex and frustrating, is the war against misinformation. Sometime in 2018, probably as a result of the often bizarre Trump presidency, fact checking became a new profession. The common phrase, referred to earlier, "You're entitled to you own opinion, but not to your own facts," almost became a mantra to some who were constantly amazed by the falsehoods coming out of the mouths of so many and in too many cases from the mouths of apparently intelligent and thoughtful persons.

If it were only a group of simple disbelievers, those negatrons, that would be bad enough, but in fact, it's a vigorous and well-organized assault on the truth. It is terribly discouraging. There is even a scary and quite disturbing website (www.checkourfact.com) which is an offshoot of the anti-science media site by the name of The Dailey Caller. Which, surprise-surprise, was founded by the science denying Fox News anchor Tucker Carlson, funded by major conservative donors such as the Koch brothers, and quoted freely by President Trump. These are terrible and destructive, and often powerful people who deny the facts, let alone seek out the truth. This could all be a fanciful joke if it weren't so outrageous.

While we have made the arguments repeatedly in support of climate change and global warming, in the following pages you'll see how so many of the climate change deniers are simply wrong, point by point.

1. We had more snow and cold weather this past winter than any time in the past 25 years.

FALSE... Globally, each of the past 10 winters have been progressively warmer. Remember also that precipitation in the form of snow is not an adequate indicator of temperature. Recent bouts of unusually cold weather in the central states in the United States are accounted for by changes in weather patterns and the jet stream moving colder temperatures to areas where they were not normally seen.

2. It hasn't gotten warmer for the past dozen years

FALSE... Nine of the past 10 summers have been the hottest that we've experienced in over 30 years

3. Even the scientists don't agree on climate change.

FALSE... Independent analysis has now shown repeatedly that 98 percent of scientists believe that climate change is real and that it is man-made.

4. These scientists can't even predict the weather. How can they know how warm it was 100 years ago or how warm it'll be in 50 years?

FALSE... There are rigorous scientific measures to determine the temperatures 100 years ago, and more. While it is true that we can't guarentee what the temperature will be in 50 years, we can make projections based on current information.

5. If the planet is getting warmer, how come some people say that the ice is thickening at the poles?

FALSE... The ice is not thickening at the poles; it's getting thinner, and in the North Pole, a lot of it actually disappears in the summer.

6. Even if the Earth is getting warmer, surely we are not to blame. We can't possibly be that powerful compared to the powers of the sun, massive volcanoes, or some other natural forces.

FALSE... We know with great certainty that the planet is warming due to the accumulation of greenhouse gases in the atmosphere arising from the using of fossil fuels as our primary source of energy. The rise in population correlates almost perfectly with the increases in the burning of fossil fuel as an energy source, the levels of the greenhouse gas CO_2 accumulating in the atmosphere, and the increase in the average global temperature.

7. We can't lower CO_2 in the atmosphere because it's needed to grow food.

FALSE... Food, plants, and all vegetation grow very well when the atmospheric CO_2 is at 300 ppm. Increasing CO_2 levels to 420 ppm where they are today has a negative impact on agriculture, including extreme weather events, causing droughts in some areas and severe flooding in others, and of course the increases in temperature.

8. We breathe in oxygen and breathe out carbon dioxide. That's a fact. Should we then stop breathing?

FALSE... And pretty silly. Of course, we breathe in O_2 and breathe out CO_2, just the reverse of plants, which breathe in CO_2 and breath out O_2.

The problem is the excessive amounts of CO_2 (e.g., greenhouse gases) that are being introduced into the atmosphere as a result of our use of fossil fuels as our major energy source. Humans produce very little CO_2 compared to other sources.

9. An increase of the Earth's temperature by a couple of degrees can't be a problem.

FALSE… It is a huge problem resulting in the melting of polar ice caps, the disappearance of mountain glaciers, rising sea levels, and a major increase in the number of extreme weather events as a result of the major changes in weather patterns.

10. We can't limit carbon dioxide emissions by limiting the use of fossil fuels, because if we do, we'll halt growth, lose jobs, and devastate the economy.

FALSE… There's every reason to expect that alternative energy sources—solar, wind, and nuclear power—will provide excellent jobs, clean energy, and allow us to function and thrive without destroying our environment. Some calculate that winning the battle against climate change will stimulate the economy, much like President Kennedy's space program in the sixties and winning World War II did in 1945.

11. Even if we could convert to 100 percent clean energy, wouldn't the planet continue to warm? Isn't it too late to do anything?

FALSE… A complete conversion to clean energy (zero carbon) would quickly stop the accumulation CO_2 in the atmosphere and put a stop to global warming. Additional promising technologies are now available to actually remove CO_2 from the atmosphere, and lowering it back down to 350 ppm could easily see us decrease global temperatures and return to the "norm" of circa 1950-1980.

In the appendix, we'll give you some instructions and tools on how to be effective in combating (only verbally) with the conspiracy theorists who embrace climate change denial.

Chapter 12
FIRST CAME DEFORESTATION THEN COMES REFORESTATION

A PENCHANT FOR CUTTING DOWN TREES

There are those who would have us believe that simply rebalancing the rates of cutting trees down with planting trees would go a long way to rectifying our climate change crises. In this chapter, we'll examine that possibility.

If we thought that the concerns about the rate at which we were cutting down trees was a recent phenomenon, we would be wrong. We might have thought that the concept of a "tree hugger" was fairly modern, maybe even a child of the sixties, but then we would be wrong as well. In fact, the very term tree hugging was first coined in 1730. A group of 294 men and 69 women in Rajasthan, then a part of India, belonging to the Bishnois branch of Hinduism, physically attached themselves to trees (hence the concept of tree hugging) in an effort to stop a group of soldiers in a small village from cutting down the trees to build a new royal palace. Many of these tree huggers were subsequently murdered—for their disobedience. But in time, their extreme actions resulted in a royal decree that outlawed the cutting of trees in the Bishnoi village, which even today stands as a small, wooded area in the middle of a desert-like landscape. The

Bishnois activity later (by 1970) inspired the Chipko ("Chipko" means "to cling" in Hindi) movement in the Himalayan hills in the north of India, where groups of women would throw their arms around trees that had been designated to be cut down. As laudable as these efforts in civil disobedience were in the interests of an important environmental issue, they were and still are little more than symbolic in a still very hostile environment.

In spite of protests being staged around the world, deforestation has continued down an increasingly aggressive path. We cut down trees for wood to build our homes and sometimes to heat them. We cut down trees to increase the land available to build highways, malls, factories, new homes, and recreational facilities. We cut down trees to increase agricultural acreage, and we cut down trees to plant palm trees to provide us with cooking oil. The reasons to cut down trees to create useable land is seemingly limitless. But we forget, or maybe just ignore, perhaps the most important role that trees play is controlling the level of greenhouse gases (CO_2) in the atmosphere. The equation is quite simple. Trees and other green plants are the primary consumers of atmospheric carbon dioxide, and cutting them down poses a problem.

1) Cut down a tree.
2) Decrease the amount of CO_2 removed from the atmosphere.
3) Increase atmospheric CO_2 and therefore the level of greenhouse gases.
4) Increase atmospheric temperature and therefore global warming.

By 1975, as populations grew and industrialization was in its major growth phase, the plights of trees and forests received little attention. But fast-forward some 25-plus years, sometime towards the turn of the millennium (2000), the consciousness of the world took notice as more concern was voiced about the quality of our environment and the potential threat of climate change. Deforestation became an international issue, and two major responses began to take hold in the thoughts, if not the actions, of many, especially the young. One was "Save a Tree" and the second was "Plant a Tree." The tree huggers and the voice of the youth, especially in the west (California, Oregon, British Columbia), began to be heard. The issue was taken up by environmentalists with genuine concerns for the future of our planet, with a special concern for our trees and forests and a new and active variant of the tree hugger was born. We would see and hear of repeated reports, many from the northwest of the United States and Canada, of so-called tree-huggers chaining themselves to trees or creating major blockages against the massive lumber companies and their primary roles in deforestation. We even saw the emergence of eco-terrorism as small numbers of environmentalists sought to use violent means to seek out their ends. While these groups made a lot of noise and received a lot of attention that continues through to today, their impact has likely been minimal. The trees and the forests continue to play second fiddle to the desires or so-called needs of an ever-increasing population. And the need to make room for agriculture, cities, highways, and industries, all at the expense of continuing to cut down trees, remains dominant.

ANOTHER INTERLUDE...BUT THIS ONE ABOUT TREES IN POETRY

So much of this book focuses on the utilitarian nature of the relationship between the boy and the tree. It is too easy to put aside the majesty of the single oak tree powerfully reaching skyward or the sheer vastness of large uninterrupted forests. In fact, throughout our history, the tree has had a very special place in the hearts of men and women, and in many ways, the trees and the forests have been romanticized. Munia Khan and Nelson Henderson wrote so elegantly in support of these programs.

Munia Khan (b.1981), a young American poet and writer puts it so simply and accurately and captures the essence of our relationship with the tree:

"Trees exhale for us so that we can inhale them to stay alive. Can we ever forget that? Let us love trees with every breath we take until we perish."

While Nelson Henderson, an old-timer, captures the existential nature of that relationship when he wrote:

"The true meaning of life is to plant trees; under whose shade you do not expect to sit."

But one of the most meaningful of simple poems about trees was penned by the American Joyce Kilmer (1866-1918):

> I think that I shall never see
> A poem lovely as a tree.
> A tree whose hungry mouth is prest
> Against the sweet earth's flowing breast;
> A tree that looks at God all day,
> And lifts her leafy arms to pray;
> A tree that may in summer wear
> A nest of robins in her hair;
> Upon whose bosom snow has lain;
> Who intimately lives with rain.
> Poems are made by fools like me,
> But only God can make a tree.

The Nature Conservancy, a global environmental non-profit organization, has taken up the mantle of putting a halt to deforestation and promoting reforestation. In 2018, they announced a Plant a Billion Trees campaign across the planet stating that, "It's a big number, but we know we can do it with your help."

Prime Minister Justin Trudeau of Canada, in an election speech, promised to go farther and plant two billion trees over a period of 10 years if he were elected. Let's stay tuned. Several leaders in the western world seemed almost giddy to compete in terms of how many trees they intended to plant.

That's all well and nice and may make us feel good, now we've finally started to at least think about doing something to combat climate change. But we also have to stop and question the numbers. How much do we have to do in order to make a difference? In a sense, it's just math and science. How many trees do we have to stop cutting down, and how many trees will we have to plant to make a difference, to put a halt to increasing levels of greenhouse gases and to put a stop to global warming and even reverse it? Unfortunately, without strong government and regulatory intervention, there is a very major disconnect between those who would **reforest** and those who would continue to **deforest.**

We get excited, and witness good press covering a program where, over a few days, young people and volunteers make a special concerted effort to plant a few million (aiming for a billion or more) trees. But think about it for a moment. How can that compete with the industrial strength deforestation being carried out by the lumber companies, by the mining companies, by the agricultural industries, and by the cutting down of trees to make way for highways, shopping centres, and new urban living areas? The reforestation programs are largely occurring in a country or a few western countries with heightened sensitivities to the realities of climate change and global warming. What about the much larger parts of world in terms of both population and land mass, where those sensitivities are absent, where their primary concerns are where their next meals are coming from and how to keep a roof over their heads?

Remember that climate change is the ultimate global issue, knowing no national borders, so in looking at both the causes of global warming and the potential solutions, global approaches are critical.

It's clear that we need a reality check and that we have to take stock. Are our many efforts to plant trees in the same ballpark as is our penchant to cut them down? In 2019, it is reported that we had a net loss of between 4 and 7.5 billion trees with at least 1.5 billion of that number coming from the continuing destruction of the Amazon rainforest. Remember the recent images of the massive fires burning in the Amazon, and for what purpose? Some say it was for grazing beef cattle, some say it was to harvest more palm trees for the worldwide demand for palm oil, and some have even suggested that it was an official government statement of climate change denial. Now how absolutely stupid would that be?

NOTE: I know that many of these issues are not always easy to follow. There is so much misinformation, and we often ask, how do we distinguish fact from fiction? That is an important objective of this book. In an appendix to this book, I've included some instructions of how to deal with or answer those **climate change deniers,** those who deny what is both clear and evident. That is, that we are in the midst of a severe, even existential, climate change crisis, and we must have a call to real action.

SO HERE IS WHERE WE–OR MORE ACCURATELY, THE TREES–STAND

If you think about it for a moment, it's really pretty simple. For many years, actually hundreds of thousands of years, we were in pretty good synchrony. We cut down lots of trees, but there were lots of trees available, so nature carried on, and we didn't think too much of it, and the impacts were minimal. Even well into the twentieth century, there was lots of coal, oil, and gas available, so we used it to warm our homes, fire our industries and so many other things. The Earth and atmosphere are collectively big places, and so all those bad emissions would simply disappear or accumulate for the most part in a fairly inobtrusive manner, it used to be said, but that was at a time long past that "dilution was the solution to pollution." If you wanted to get rid of something, just throw it into the water or burn it and let it escape into the atmosphere. Then somehow, maybe around 50 years ago, when people started to see what was going on around and began to listen, it was as if Mother Earth was shouting at us, and this is what we imagine her saying:

"Enough already, for all these years you've been insulting me and taking advantage of my good nature (pun intended), allowing me to take care of your continued mess. I can't take it any longer. I'm at my wits end, and from now

on, you're going to have to change if you plan to continue to live and thrive here on Earth."

Another way of looking at it was spelled out on a billboard next to a highway:

**THE INDUSTRIAL REVOLUTION BROUGHT THE DEVELOPED WORLD
150 YEARS OF UNPRECEDENTED PROSPERITY
GLOBAL WARMING IS THE BILL
THE BILL THAT HAS NOW COME DUE**

SO LET'S JUST PLANT TREES

It was the year 2018, and the boy and the tree were growing more and more despondent. It was now perfectly clear that they were traveling down a road in the wrong direction and, heaven forbid, on a one-way street. Trees were disappearing, and the boy's asthma was worsening. The boy, then a man, had become a farmer living in India. But his coastal farm could no longer support any form of agriculture as a result of sea (salt) water flooding all of his fields, and his future looked bleak. He moved back into a crowded, polluted city far from his home and his farm. He was depressed. Then suddenly a lightbulb turned on in his head, and he said to himself and to the tree:

"Why don't we simply plant more trees? That would solve all our problems. And we could use some more greenery, some more parks. Let's just start by planting a few trees."

The trees need CO_2 to grow (remember photosynthesis), and God knows we have plenty of it, so they'll get the CO_2 from the atmosphere and stop its accumulation, decrease the amount of greenhouse gas in the atmosphere, and therefore stop global warming. In fact, if we were able to plant enough trees, we could even remove enough carbon dioxide from the atmosphere to actually reverse global warming and solve all—well, not all, but many, of our climate-based problems.

By the way, we weren't the only ones to come up with this simple, brilliant idea. The Nature Conservancy committed to planting a billion trees. Governments across the world jumped on board—well, sort of—and it became a movement with youth in classrooms taking up the cause. China initially somewhat surprisingly announced its plans to plant over 10 billion trees a year for the foreseeable future. Their commitment appears to have been very sincere in their response to huge environment stresses. Remember the disastrous air quality during the Beijing Olympics? And it appears that they're sticking to their commitments. The Chinese also recognized the dramatic impact of global warming in creating large desert-like conditions in large parts of its central plains, which less than a century ago were relatively large, fertile grasslands. The commitments of Canada (two billion trees over 10 years) and the United States (3.3 billion trees annually) are all exciting initiatives.

The fact is, however, that it's not clear whether these commitments are sufficient or even come close to reversing the problems that we've created and even more so how serious these governments really are. Certainly, the pace thus far has not even come close to what the needs of a real reforestation program would encompass. In 1986, the American environmentalist Jay Westerveld coined the word **"Greenwashing."** He used the word to describe hotels who encouraged their guests to reuse their towels as part of a wider environmental strategy when it was really only meant as a cost-saving measure. At the time, the hotels that promoted this "reuse" actually lacked any environmental initiatives. It was simply a ruse, and yes, it does have elements in common with brainwashing. In recent years, greenwashing has grown to refer to people or companies and even governments who attempt to mislead us with environmental claims that are really meant to take attention away from the actual actions of the individual, company, or government. It's analogous to whitewashing, as in when a company conveys a false impression that its products are environmentally sound when they're not. You will all likely have seen TV-based Green Energy claims put forward by some of the major multinational oil and gas companies. They spend less than one-tenth of one percent of their research budgets on environmentally friendly projects and expect people to believe that they are committed to carbon-neutral energy sources. That's greenwashing: It's wrong; too many people are falling for it; and it's a major disservice to us all.

Taken at face value, the reforestation programs should be pretty good news. Let's pause and ask how many trees would have to be planted and who would plant them and where would they be planted, and if there is even enough land to plant them, and of course, how much is it going to cost? And how long would they take to grow so that they would start to make a difference? These are critical questions that have to be addressed if we intend to stake our futures in an important way in the hands of reforestation. Put another and perhaps more optimistic and realistic way, we should ask: What part of the pie of "stopping and reversing the progression of global warming" can we actually assign to the process of reforestation? Let's revisit the numbers to see if reforestation represents a rational solution to our climate change crisis.

FIG 16 REFORESTATION...PLANTING TREES
(Fixing climate change, one tree at a time.)

But first lets make sure that we aren't perceived as recycling, reforestation luddites; that we're not belittling or "rubbing salt into a wound" of the many initiatives that concerned individuals are taking to save the plant. Every action counts, and I would be terribly remiss if I dismissed any attempts, no matter how small, that were taken on by individuals attempting to make a difference. While the problems are certainly enormous in scale, everyone has to take responsibility; the notion that you have to think globally but act locally is essential. Every tree planted is meaningful. Every plastic bottle or plastic bag not purchased can make a difference. Every time you save energy by walking and not driving, by keeping the heat down in your house, selling your gas guzzler and purchasing an electric car—or even better, a bicycle—all contribute to combatting the enemy, which is global warming. Every effort, no matter how small, contributes. Every effort also counts for spreading the word and increasing awareness.

But while those efforts taken on by individuals or even by communities are significant, they are small in comparison to the enormity of the problem, which has to be undertaken on a global scale. One cannot simply adopt one niche of the problem (for instance, refuse to use plastic bags) and assume that you're doing your thing, making your contribution. You are; however, it's not enough. The issue of global warming is massive, and the steps we take to solve the problem have to be both substantial and sensible.

I also want to warn you not to allow yourself to be "greenwashed."

So let's get back to the issue of planting trees and see how it might work. The question we're asking here is whether we can plant enough trees so that enough carbon dioxide will be pulled out of the atmosphere to not only stop the continued progression of global warming but actually to reverse it at least a bit. Say we want to decrease the atmospheric levels of CO_2 from 419 ppm, which it currently, is to 350 ppm, which is the level it was in 1985. Although an arbitrary date, it appears to be an inflection point where things started to get really bad in terms of climate change. If we could go back to that level of CO_2, we might avert many of the crises that are looming.

Continuing this thought experiment, if we deal in tons of carbon and give ourselves a deadline of 2032 to make that reduction of atmospheric CO2 to 350 ppm. We can reframe the question to ask: How many trees do we have to plant in order to remove the necessary tons of carbon dioxide from the atmosphere over the next 10 years?

One ppm of carbon dioxide in the atmosphere represents 2.1 billion metric tons of carbon. A metric ton is 1,000 kilograms. While the nomenclature is a bit confusing, that's what is commonly used, so we'll have to stick to that.

Therefore, if we want to reduce the atmospheric CO_2 by 70 ppm (the rounded off difference between 417 and 350), we have to remove 147 billion metric tons of CO_2 from the atmosphere, and we have five years to do it, so in rounded terms the annual number is 30 billion metric tons (or 30 trillion kilograms) of carbon each year.

Can it be done? Let's try to do the math, one tree at a time.

An average tree of an average age in an average climate can remove up to 20 kilograms of carbon dioxide from the atmosphere in a year; 30 trillion kilograms of carbon, which divided by 20 is equal to almost 1.5 trillion trees.

What do 1.5 trillion trees look like? Well, since it is estimated that there are already 3 trillion trees, that means we have to increase the current number of trees by 50 percent to 4.5 trillion trees, and note: That is without cutting anymore down. What would that look like?

An average, healthy forest might have as many as 500 mature trees per acre, so to plant 1.5 trillion new trees, you would need 3 billion acres or approximately 4.5 million square miles of forest. There are currently about 10 billion acres of forest in the world, down about 60 percent from pre-industrial times.

Don't be intimidated by the size of the numbers; they'll make sense in just a moment. The Earth's surface area is reported as 196.7 million square miles, of which 29.2 percent or 57.5 million square miles is land. Therefore, the 4.5 million square miles (the land required to plant 1.5 trillion trees) is less than 10 percent, so although it's a huge number, it's really not that preposterous. Take the major countries with land to spare (Canada, Brazil, Soviet Union, China, and a few others), and one might be able to get the job done. If only it would (not wood) be that easy. Let's examine for a moment some of the limitations.

1) We're still dealing with an enormous area, since 4.5 million square miles is more than the total land area of the lower 48 states of the United States or all of Western Europe or 60 percent of the arable land mass of Canada.
2) Remember that there are huge land masses around the world that are not hospitable to tree growth (i.e., the Arctic and Antarctica). Increasingly large areas of Asia, Africa, and the Americas that are more desert-like than ever make major tree-planting efforts implausible. So, the notion that you would only have to reforest 6 percent of the total landmass is a serious underestimate.
3) You can't expect a newly planted tree to "do its stuff" right off the bat. In some areas, like the Amazon, the trees may mature in as little as 10 years, whereas in other areas, such as in the north where winters are cold and dark, trees may take more than 50 years to mature and achieve their maximum carbon dioxide usage.
4) And how are going to going to get the world's commitment to stop cutting down trees for the multitude of uses described earlier? So the news is not entirely good. Even if we could plant the additional 1.5 trillion trees to lower atmospheric CO_2, it would probably fall well short of our needs and certainly too late for our timeframe. The world would have suffered too much additional damage from global warming before the reforestation solutions actually kicked in.

But these calculations do tell us just how important reforestation is and how a portion of the problem could be solved by strong reforestation programs. In fact, a combination of reforestation and complete conversion from burning fossil fuels over the next 10 years would be effective in lowering atmospheric CO_2 by the necessary 70 ppm, and we would have licked the global warming impact in our lifetimes. The strict caveat in this scenario is that we would have to stop using coal, oil, and gas as our energy source effective immediately. That is within a couple of years and not by the year 2050 as so many of our politicians and leaders who actually believe in the perils of climate change are suggesting. Any of the various solutions to the climate change crisis require good reality checks. We'll deal with possible solutions in a subsequent chapter but:

WOULDN'T IT BE FANTASTIC IF OUR POLITICAL AND INDUSTRIAL LEADERS HAD THE UNDERSTANDING AND THE WILL TO DO JUST THAT AND ACT WITHIN THE TIME FRAME REQUIRED?

In essence, we're in a foot race between the forces that continue to increase atmospheric CO_2 and those that would decrease it.

<u>INCREASE</u>	<u>DECREASE</u>
Population Increase	Alternative non-CO_2 emitting energy sources (e.g. solar, wind, nuclear)*
Industrialization	Carbon taxes/conservation
Use more oil and gas	Carbon Capture/Reuse*
Cut down more trees	Reforestation
Increase food production	Cellular Agriculture*

- These are all very exciting technologies that will be described in a subsequent chapter dealing with solutions.

Chapter 13
CLIMATE CHANGE AND THE BOY'S HEALTH

For so many years, the tree and I were just good buddies. We only did good for one another. Well, if I do some honest soul searching, I can see that really the relationship was pretty one directional. The tree provided me with seemingly unlimited benefits in terms of warmth, shelter, and fuel, to say nothing of food. And if the truth be told, I'm guilty of taking enormous advantage of the relationship, and now that misuse is coming back to bite me…big time. And to put it bluntly and literally, it's really started to become "the death of me."

It is probably not entirely obvious to the average person living under high standards in well developed countries in the west that the impact of climate change on human health is enormous. That, in fact, it's getting worse year by year. So, let's look at the various parameters of climate change and their impact on our health.

Let's start with the most common elements of climate change and they are:

Increasing CO_2 levels.
Rising temperatures.
More extreme weather.

Rising sea levels.
Increased greenhouse gases.
Increased noxious elements entering our air and water.

The impacts are straightforward when you begin to dissect them.

Rising temperatures give rise to extreme heat, which results in heat related illness and death, especially due to cardiovascular failure. We see annual pictures of such impact, especially in the areas of Sub-Saharan Africa, and areas in Asia and South America. You can recall a summer in the late 2010s, when a heat wave killed 35,000 people in Europe. In Paris alone, 15,000 people died, mostly the elderly who had been left to their own devices when a massive heat wave struck. It was a devastating time, as many elderly people died of hyperthermia at a time when their children were vacationing outside of the city.

You don't have to be frail and elderly or suffer hyperthermia for climate change to have a serious impact on your health. Global warming can have a consequential impact on respiratory health (i.e. your lungs and how and what you breathe). The increased CO_2 along with the increased temperatures exacerbate a whole host of conditions, such as asthma and cardiovascular stress, including Chronic Obstructive Pulmonary Disease (COPD).

Rising temperatures also give rise to severe weather events, which result in injuries, fatalities, and often serious impacts on mental health. We don't have to travel the world to observe this. We can just look at the impact in the United States. The California fires of 2018, 2019, and 2020 were devastating, claiming lives and untold damages. Hurricane Katrina flooded more than 80 percent of New Orleans and resulted in over 1,500 deaths with damage estimated at over $70 billion. Some of the worst hurricane seasons ever have occurred over the past few years, creating massive death and destruction throughout the Caribbean and causing many tens of billions of dollars of damage in the southeastern United States. There have been reports of major increases in the incidents of mental illness associated with severe weather events. Recently we witnessed a huge polar vortex covering much of Texas with week-long power outages, freezing temperatures, and much of the state was also without water as water-mains burst. The cost of just a single episode of cold weather in a single state cost in excess of $100 billion.

Now, wait a minute. How can it be that global warming causes freezing weather in the southern United States? It does, and it's a good and important question, which we'll address in Chapter 15.

Major extreme weather events have led to increasing issues in air pollution as air quality decreases due to smog, increased ozone, and particles in the air due to wildfires. All of these are serious lung irritants resulting in subsequent increases in chronic respiratory disease (CRD), asthma and cardiovascular disease, especially COPD (Chronic Obstructive Pulmonary Disease). Over the time period from 1980 to 2015, there was an 87 percent increase in incidents of COPD in the United States with similar and even more dramatic rises in other parts of the world. Similarly, the incidents of childhood asthma have more than doubled over that period of time. While these changes correlate on a time basis with global warming, it is probably the quality of the air, the presence of pollutants resulting from the excessive burning of fossil fuels that are the major culprits.

A recent study from Harvard University reported that more than 10 million people are dying each and every year as a result of breathing air contaminated with pollutants resulting from burning fossil fuels. Incredibly, that's the leading cause of premature death in poorer parts of India and China, but the numbers in the United States and Canada are also significant.

FIG. 17 CLIMATE CHANGE REFUGEES ESCAPING NORTH AFRICA

More extreme weather and changes in local weather and temperatures overtime also results in a potentially devastating phenomenon referred to as <u>changes in vector ecology.</u> Simply put, vector ecology refers to the study of disease-bearing organisms/insects, their behaviours, and environments. In this case, we're talking about pathogens infecting humans, but it could be spread to other animals or plants as well. A good example is the mosquito that carries the malaria virus. Vector ecology refers to the habitat or area where those vectors "live" and therefore where they might infect humans. Climate change and especially global warming can change the ecology of that vector/insect and therefore change where that pathogen or virus might be carried and where it might infect humans. We now see different diseases with those vectors moving north from the equatorial regions in South America, and all of a sudden, diseases that were never found even in the warmer climates of Florida and Georgia start to appear. In this case, the vector went north along with the changing weather patterns. A very major concern is that persons living further north may never have come in contact with the particular vector or pathogen, and so are "naïve" to it with potentially very serious implications. Some of these diseases include viruses such as Dengue, West Nile, Zika, and others, some of which are terribly lethal to naïve populations.

There is an equivalent to vector ecology that is seen in various areas of South America and Sub-Saharan Africa, where humans can actually function as the vectors as they flee certain equatorial regions in South America and Sub-Saharan Africa for "greener" pastures in response to famine and drought conditions in their countries. This is another direct response to climate change and global warming with enormous parts of the Earth being devastated by heat and drought, making huge areas uninhabitable. We have only to remember the images of refugees from Africa fleeing into the Mediterranean towards Europe in rubber dinghies, trying to escape desperate political and financial conditions where climate change and global warming have made it nearly impossible for them to feed their families.

Extreme weather, pollution, and rising sea levels can result in increasing allergens in the atmosphere and again the development of respiratory allergies and asthma as well as added pressure on our cardiovascular systems. As a result of weather changes, all sorts of reasons can cause the time and quantity of increased pollen and allergens in the atmosphere, including mold and even poison ivy.

Rising sea levels will undoubtedly directly impact the quality of the water that we use individually or in our food preparation. Think for a moment of the flooding that occurred in New Orleans in 2005, or the frequency of low-lying coastal areas in India and in eastern Africa being flooded by ocean waters. Saltwater contamination has destroyed tens of millions of acres of low-lying agricultural lands, rendering them infertile. We often see the spread of a whole host of devastating infectious diseases, with names that are difficult to spell and even harder to pronounce. Diseases such as cholera, cryptosporidiosis, campylobacter, leptospirosis, harmful algal blooms, and other viral pathogens that are increasingly being transmitted to the human population as a result of the decrease in water quality and the changes in vector ecology. Increased CO_2 resulting in global warming can also have huge impacts on water and food supplies, resulting in widespread malnutrition and diarrhea.

And finally, the increase in atmospheric CO_2 can result in many extremes of en-vironmental degradation, resulting in civil conflict, forced migration, and the subsequent impact on mental health. Again, consider the millions of people around the globe who are leaving their homes in utter distress as a result of severe challenges in food safety and food security as they seek to move, usually north, to what they hope to be more friendly environments. More often than not, the travel itself results in huge hardships and very often death from the perilous journey. They seek refuge in northern Europe and, when possible, in the United States and Canada. Many of these refugees have been labelled as political refugees or economic refugees, but really, they might better be described as climate change refugees. In many, if not most, instances it was the issue of changes in climate and changes in food security that over quite short periods of time people were forced out of their homes or countries. The notion of climate change refugees is most prevalent in the migration of those impoverished and desperate from Sub-Saharan Africa and parts of South America, those who are fleeing heat, drought, and oppression. When they arrive at their destinations, their lives are usually far more difficult than that they anticipated.

A recurring theme around the impacts of climate change and global warming is in the attempt to "monetize" the costs of these changes on human health. Those are the true costs of global warming. It's of interest of course to compare that number to what it might cost to put a halt to and even reverse global warming. The World Health Organization (WHO) has indeed taken a crack at those calculations.

So, let's take a few minutes to look at those arguments and the associated numbers or costs.

But first, a word from the World Health Organization and their section on Public Health, Environmental and Social Determinants of Health. They put forth a fairly simple and straightforward statement:

"The true cost of climate change is felt in our hospitals and in our lungs. The health burden of polluting energy sources is now so high, that moving to cleaner and more sustainable choices for energy supply, transport and food systems effectively pays for itself."

In fact, they add: "When health is taken into account, climate change is an opportunity, not a cost."

That's actually a very powerful argument, but it too often falls on the deaf ears of our industrial leaders, who are more concerned about their quarterly reports, or our political leaders, who are more concerned about the next election cycle. But it does harken back to the saying that "an ounce of prevention is worth a pound of cure." It is difficult to understand why our political and industrial leaders don't get it. Practicing a precaution or prevention in advance of a crisis occurring is always preferable to having to fix up or clean up after the crisis has occurred. It seems pretty straightforward.

Making sense of the financial -numbers around climate change is a bit futile, however, it's still worthwhile to look at some of the numbers that seem to have a level of credence amongst many of the people who develop these calculations. The following facts come from the United Nations and the World Health Organization.

1) Exposure to air pollution causes 10.5 million deaths annually and $5.1 trillion in losses globally.
2) Within the next 20 years, an additional 290 million people will be suffering from and die as a result of malnutrition.
3) In the 15 countries that emit the most greenhouse gases, the direct health impact of air pollution is estimated at four percent of their GDP, or $2.8 trillion annually. It's really striking to read that the estimated costs to meet the standards set by the Paris Accord would only cost one percent of the global GDP, or about $1.5 trillion per year, between now and 2050.

4) The economic impact (loss in GDP) of global warming is felt most dramatically in poorer African, Asian, and South American countries that sit close to the equator. No surprise that these are also the countries with less clout to effect change.

5) The amount of money that countries with well developed economies spend subsidizing the fossil fuel industries, especially oil and gas, could easily be used for a complete conversion of the globe's energy utilization to largely renewable sources, mainly solar. Or it could be wind, nuclear, hydrogen, or other renewable non-fossil fuel-based energy sources. We'll cover these topics in Chapter 17. **What a simple idea, and the barriers are entirely political and not economic.** In fact, following such a conversion, economic growth would be substantial, and the health impacts of climate change would be curtailed and even reversed if we could reduce atmospheric carbon dioxide. And that too is doable.

Chapter 14

THE COSTS OF CLIMATE CHANGE

On the one hand, it would be nice to be able to simply say we're in crisis; humanity is in crisis; the crisis is an existential threat; so just get on with it and do what's required. Think about the reaction of President Franklin Roosevelt and the Americans in response to the war in Europe, the existential threat that the Nazis posed, and the American decision to go all in and help Britain and the Allies to win the war. There was little consideration of the costs; there was first and foremost a need and the will to get the job done. If the world, and especially the developed nations, had the will and the understanding of the existential threat that global warming and climate change represent, their actions would be similar to that of President Roosevelt and the American people in 1941. But one doesn't have to time travel back 70 years to see how resources suddenly appear when the need is properly demonstrated. The response of the United States, Canada, and Western European countries to the COVID-19 pandemic has similarly been an all-in response. They had committed in excess of $10 trillion even before the pandemic was over. The equivalent of those amounts spent on protecting the population, especially in North America, from the virus would go a long way towards solving the climate change crisis. If only our leaders could understand and acknowledge that the climate change crisis is every bit as serious as the COVID-19 pandemic.

That's simply what you do when faced with an existential crisis...but first you have to understand and acknowledge the crisis.

In spite of the increased talk and attention, our political, financial, and industrial leaders are not yet in crisis mode when it comes to climate change. It is important that we at least examine what the actual costs of climate change are, and what would we have to spend in order reverse it That's a difficult undertaking and sort of a dog's breakfast (for any millennial readers, that's an old way of saying a messy situation) in determining what's included and what's not; what are the costs of not responding, and what are the costs of making the necessary changes.

A simplified version of the question would be to ask what would be the near-term costs of shutting down the fossil fuel business and what would be the replacement costs, not only the fixed asset costs, but the social and economic costs to the people engaged in those industries?

The predicament in making those arguments evokes a number of important quotes:

One by the famous New York Baseball baseball Manager Casey Stengel:
"Never Make Predictions, Especially about the Future"

And the second by Peter Drucker, the famous American management consultant, thinker, and author, who prominently said:
"The best way to predict the future is to create it."

And if we take the liberty to tweak that quote slightly to reflect the critical threads running through this book:
"The best way to assure the future is to clean the Earth."

Let's start with the issue of how much climate change or global warming actually costs society in real money, realizing of course that climate change is much more than just economics. How do you monetize the extreme anxiety that young people are exhibiting in contemplating their futures (eco-anxiety), or the trauma (quite aside from the direct damage) that flood victims in India, the Caribbean, or Florida exhibit as a result of extreme weather events? So point made. This is a very complex set of considerations

Initially at the onset of writing this book, we wondered what responses these numbers might elicit. What does $10 trillion even mean? Readers would question the validity of the numbers. Where do they come from? Are they being exaggerated to make a point or, heaven forbid, to sell more copies of a book? Let's look at some of the numbers, some of the facts. But actually, these arguments have been made easier, unfortunately by the COVID-19 pandemic. Even before the end of the pandemic, the United States government alone had come up with almost 10 trillion dollars to stop the social and economic calamity associated with the pandemic. And this was on top of the tens of billions of dollars dedicated to research, development, and the production of new therapies and vaccines. The question of where the money will come from was hardly ever discussed. Newt Gingrich, ultra conservative commentator and former Speaker of the House of Representatives in the United States, said it simply: **"This is a crisis. We'll deal with the costs when it's over."**

In the past 20 years, extreme weather has cost the United States $1.6 trillion in direct costs; the wildfires in California in 2018 costs insurance companies $24 billion, and the hurricanes in the Caribbean and southeast part of the Unites States another $35 billion. Between 2016 and 2018, the direct costs of extreme weather in the United States averaged $150 billion per year. By all accounts, between the wildfires across the west and the hurricanes in the southeast in 2020, the total costs will be well in excess of $250 billion. What's especially alarming is that these numbers appear to be increasing every year, directly related to the increase in global warming and the incidents of extreme weather events.

The numbers from the United States are being used here because they happen to be reliable and accessible. But think about how the number should be multiplied by some factor to get worldwide numbers. Suffice it to say that the impacts and costs are enormous. In an eerily tragic way, in less well-developed regions of the world, the devastating effects of extreme weather events cannot be monetized because the changes are permanent. There is no money to rebuild, and people suffer, and when they can, they simply move on—or at the very least attempt to. You don't have to go very far offshore of North America to see incredible and permanent damage. The 2010 earthquake in Haiti, one of the western hemisphere's poorest countries, was horrific in terms of loss of life and physical damage, and to date, the repairs have been minimal. Even in Puerto Rico, a part of the United States, the reconstruction of major parts of the island following Maria, a Category 5 hurricane, has been minimal, and many people have been forced to relocate.

In other areas of the world, Sub-Saharan Africa, India, parts of South America, and in the central plains of China, climate change has led to mass migration. Immigrants are forced to leave flooded coastlines and drought-stricken farmland for safety, solely in the effort to find a place to house and feed their families. In the past 10 years alone, 50 million people (excluding those who have been forced to leave millions of hectares of drought-stricken agricultural land in China) have been displaced, and that might just be the start. In China, that number is closer to 200 million, although exact figures are difficult to obtain.

As CO_2 in the atmosphere increases, a lot of the excess is absorbed into the oceans, which have become 30 percent more acidic since 1950, resulting in mass extinction of many species of marine life, including the death of 50 percent of the world's coral reefs over the past 30 years. This in turn has resulted in untold numbers of fishermen who, in effect, have lost their means of earning a living. The economic costs to wild fisheries and aquaculture due to ocean acidification by 2020 had reached $300 billion per annum. And this doesn't take into consideration the increased costs to consumers of a diminishing resource.

The World Health Organization states that exposure to air pollution causes 10 million deaths worldwide per year at a cost of over $5 trillion. The numbers are often difficult to fathom and often impossible to compare, but the general concepts and the degrees are clear.

It has been suggested we might save the world $10 trillion per year in direct climate change costs just by the year 2050 if we can keep the extent of global warming to 1.5 degrees Celsius. The current trajectory of global warming would see the temperature rise by as much as 3.5 degrees Celsius by that year, assuming we do nothing to stop that momentum and mitigate the impact of climate change. If we do nothing to curb our reckless dependence on burning fossil fuels as a source of energy, we will face catastrophic conditions, which some have called a "doomsday scenario," described depressingly in a subsequent chapter. It's hardly meaningful to put a dollar figure to our very extinction.

These are all stunningly large numbers. Some of them represent real hard numbers (e.g. costs to rebuild cities or islands ravaged by Category 5 hurricanes, or loss of arable lands due to coastal flooding, or decreases in worldwide fisheries, direct health costs), while others might be considered softer numbers (e.g. potential years lost due to disease, lost opportunities due to climate change).

The numbers are not only huge, but they also overshadow the real costs of fixing the problem. Following on the calculations of the WHO:

> *In the 15 countries that emit the most greenhouse gases (USA, China, Japan, Germany, France etc.), the health impacts of air pollution are estimated to cost more than 4% of their GDP or $2.4 Trillion/year. That's about twice the cost that actions to meet the Paris Climate Accord would require or $1 Trillion/year over a ten-year period.*

The takeaway messages are pretty clear.

1) The costs associated with the damage that climate change is producing in most everything that we do are overwhelming.
2) There is a very good consensus that the damage is continuing to occur at accelerating rates. Failing to come to grips with many of the impacts on climate change will continue to cost us and make our lives more miserable each and every year. This may finally lead to the existential question, and that is humanity's survival or extinction.
3) The costs to fix the problem (e.g. stop burning ALL fossil fuels, recapture carbon from the atmosphere, reforest the Amazon and large parts of Africa) are huge. But they are certainly not out of line with the amounts of money we were quick to find to "fix" other catastrophic events: For example, the American response to World War II in Europe in 1941 and the western countries' responses to the COVID-19 pandemic beginning in 2020.
4) The greater costs will be political as we impose the appropriate carbon tax in the short term, tax the fossil fuel industry out of existence, and force the food industries to implement entirely new strategies with respect to the use of land, water, and energy. Those are just a few of the people who need to adapt and change.
5) And in the final analysis, the planet (and that includes all of us) will be both HEALTHIER and WEALTHIER.

And now for the number that should surely cause you to immediately ask, "Well why don't we just do it?" and "What's the bottom line?" The cost to transform the entire United States energy grid from burning fossil fuels to renewable energy is just $4.5 trillion. That's a fraction, even a small fraction, of the current costs that we incur dealing with the harm of climate change. We'll talk about that more in our "solutions" chapter.

Chapter 15
GLOBAL WARMING IS A FACT

In essence, this really is a book about climate change and a not so subtle warning of the crises that humanity is currently facing as a result of a significant global warming. It also provides a recipe for how we can extract ourselves from this "hot mess," and it's the extension of the story about the boy and the tree and what's happened as the relationship crossed boundaries and the boy became much too dependent on the tree. By the end of the twentieth century, we had become almost entirely dependent on burning fossil fuel, either directly or indirectly as a source of energy for virtually all of what we do. Global warming and climate change are the result. It would be nice if we didn't have to write to convince you that global warming is real and a very serious threat. After all, 98 percent of scientists who work in the area believe that not only is global warming real, but that it is manmade.

But regardless of the facts, any discussion of climate change still generates great controversy. The issues of change in day-to-day weather or even month-to-month weather is not that easy for any individual to either monitor or comprehend. We all experience that "coldest day ever" or that "warmest day ever," or that's "the most rain I've ever experienced," but it's much harder for us to relate to year-over-year changes in the climate, let alone global warming, which after all only amounts to a couple of degrees over decades. And then we have "authoritative" folks telling us that it's not anywhere near as bad as we are being told and that the scientists and those who talk about climate change with passion and concern and who quite frankly have the numbers and the science on their side are really "prophets of doom." That was the term former President Trump used to infer that climate change activists are really a bunch of left-wing activists, socialists, or worse. The boogeyman in the room is always the economy as the guilting question is always there: Do you have any idea how much it would cost to wean ourselves off fossil fuels, and are there really any possible alternatives? The discussion summons the quote by Derek Bok, former president of Harvard University:

"If you think education is expensive, try ignorance."

The corollary being is that if you think it's expensive to take on global warming today, think about what the price will be if we don't, and how the price will increase sharply with time.

But surely, it's not that simple. It can't just be settled by a debate of "he says/she says," fighting in a sandbox...or in the snow or rain.

So why are there so many naysayers? Why are there so many forces espousing the notion that global warming is a hoax or the workings of doomsday naysayers or left-wing conspirators? The answer is probably pretty straightforward. The former secretary of the United States Treasury, Steven Mnuchin's response to Scandinavian climate change activist Greta Thunberg is both telling and disgusting as he tells her to go to university and to study economics before instructing the world on the dangers of global warming. This is unfortunately much too common a response. It reflects Steven Mnuchin's, and many other political and industrial leaders' complete ignorance of issues of climate change. They seem singularly concerned with the short-term interests of their constituent's pocketbooks, and quite frankly their own pocketbooks, all probably well lined with the profits derived from activities of the fossil fuel industries. The worldwide investment in the fossil fuel business is massive. Whole countries and not only those in the Middle East are dependent on fossil fuels; some of the world's largest companies are based either directly or indirectly on fossil fuels. It's not only the fossil fuel industries. It's all the financial institutions and all other industries (e.g., automotive, transport, manufacturing, agriculture, etc.). Then in turn, many of our political leaders, knowing where their bread is buttered, are also very strongly vested in the continued success of the fossil fuel industry. And note that countries such as the United States and Canada are probably as compromised as countries in the Middle East.

If you take the world's 20 largest international corporations and subtract out the six companies that are primarily IT companies (i.e., Apple, Microsoft, Google, Facebook. etc.) then 14 of the others, or 100 percent of the other top public companies, are directly and heavily invested in fossil fuels; either oil and gas companies, automotive companies, or financial institutions which are heavily invested in both. Make no mistake about it, the costs to stem the tide of global warming and indeed to reverse it will be enormous. And these 14 companies will be the hardest hit unless they heed the wake-up call and make serious adjustments to their business plans. But even more so, the cost of not taking on climate change and succeeding in "slaying the dragon" will be even greater both from a short and long-term economic perspective.

The level of drama here is without hesitation. The arguments laid out in this book are based on facts, and critical action is required if we are to save humanity as we know it. Given the short-term views of either the stock market, or the election intervals, it is disappointing but perhaps not that surprising that neither the private sector nor the political process are ready or willing to accept the direct link between global warming and the fossil fuel business. It's simply bad for business. But it also points to incredibly weak leadership. Note that even the search for energy alternatives (e.g., solar, wind, nuclear) are supported with lukewarm efforts, if at all, by the vast majority of our industrial, financial, and political leadership. It is noteworthy that a number of industries have stepped up to the plate to anticipate that we may have to move to a state that is much less dependent on fossil fuels and more likely to their complete banning. But be warned not to allow yourself to be greenwashed. Please don't be fooled by those companies who spend more money marketing themselves as green than they do investing in alternative green energy and practices.

FIG 18 WALL STREET, FINANCIAL INSTITUTIONS, FOSSIL FUELS
(Herein lies the source of both the problem—and the solution.)

Putting on an optimistic face, it's also interesting and somewhat heartening that there are some corporations that are a step and a half ahead of our politicians and the fossil fuel producers/users. In a very real sense, they're hedging their bets and questioning what a fossil fuel energy-free economy might look like and how it might actually indeed be good for business. Think for a moment of the success of the Tesla automobiles and the investments that many of traditional combustion engine car companies are putting into development of EVs, or electric vehicles. Even the massive fossil fuel companies such as Exxon Mobil and BP are investing heavily in renewable energy sources like solar, wind, and nuclear. It's still hard to see how serious they are, but even so, it does indicate some progress. It's not easy to determine how serious they are or whether this is just an additional effort at greenwashing.

In earlier chapters, we examined the progression of this "problem" following the life of the boy and the life of the tree. While the seeds of conflict were planted when man first started burning wood or coal, the actual conflict, or the impact of that conflict, only started to become obvious after 1950. The seriousness of the conflict might have been seen as early as the second Industrial Revolution in Britain (circa 1850), when steam driven engines drove the industries and large quantities of coal were the source of energy. As early as that time, we also started to see the health hazards of mining and using coal, as the first reports of black lung disease and other devastating health impacts were evident. But from the perspective of global warming, things only really started to change and accelerate following World War II. Population growth, industrialization, and the expansion of the fossil fuel industry as an energy source driving industry all started to sharply expand. During that period, only a very few people were warning about the dangers of excessive burning of fossil fuels and the implications for the world's climate. But by the year 2000, the proverbial cat was out of the bag, and serious alarms began to be heard as the impacts of climate change started to become evident, and many viewed it as a serious problem. By then, things had started to move very quickly as we entered dramatically into this modern age of acceleration. (Read Thomas Friedman and his book: *Thank You for Being Late: An Optimist's Guide to Thriving in the Age of Acceleration*.) By the year 2020, the evidence is overwhelming that we are now experiencing a series of linked phenomena.

What are the crises?
In the recent past and especially since the start of the new millennium, there are 10 items that are crystal clear.

1) Population growth has continued almost unabated, and it appears that at least globally only disease and catastrophic events have been able to slow it down.

2) Fossil fuel usage continues to increase at almost the same rate as the population.

3) Greenhouse gases, including carbon dioxide, have increased dramatically effectively in parallel with the burning of fossil fuels.

4) Global temperature, with few exceptions, has gone up year by year, and in eight out of the past 10 years, we have recorded highest annual temperatures.

5) Drought and the devastation on agriculture in large parts of Africa and Asia are becoming much more frequent.

6) The temperatures in our oceans are increasing, and the waters are becoming more acidic with devastating effects on marine life.

7) Polar ice caps are melting, and glaciers are retreating at alarming rates.

8) Ocean and sea levels are rising.

9) Extreme weather events are occurring much more frequently in all parts of the world.

10) Global deforestation continues at an alarming rate in spite of tree planting initiatives.

A direct result of one or more of these crises also includes major periods and events of social unrest, upheaval, and displacement. We have only to point to whole societies and economies in Sub-Saharan Africa, which have been devastated by drought and starvation resulting in the flood of economic, social, and political refugees north into Europe. The concept of climate change refugees has become very real. And the phenomenon is not limited to North

Africa. We witness the same phenomenon repeated in various parts of Asia, and Central and South America.

So there's no confusion, let's repeat, or rather review for a moment, what we mean by climate change and global warming. Are the terms synonymous, and when are they used? These definitions are largely taken from the communication team at NASA's Jet Propulsion Laboratory at the California Institute of Technology. First a definition of climate and weather:

CLIMATE refers to the long-term regional or even global average of temperature, humidity, and rainfall patterns over seasons, years, or even decades. As an example, we might say that this area has traditionally over a period of time experienced a lot of rain. Or this area is generally thought of as being very dry or very humid. So, we might say that a particular area has a dry or a humid climate.

WEATHER, on the other hand, refers to the actual atmospheric conditions that occur locally over short periods of time as in minutes, hours, days, and sometimes even longer. Here, we're referring to temperature, humidity, precipitation, wind conditions, etc. As a quick reminder – please refer to Mark Twain's quote:

> *"If you don't like the weather in New England now, just wait a few minutes."*

CLIMATE CHANGE usually refers to the long term (measured in decades, centuries, or even millennia) shift in global or regional climates or climate patterns that have occurred largely, but not exclusively, as a result of increases in global average temperatures. These changes may also be local, as the temperature or precipitation in a region having changed over a period of time.

GLOBAL WARMING generally refers to the long term (in years or decades, not months or centuries) warming of the planet since the late nineteenth or early twentieth centuries, largely coinciding with the development of the industrial revolution and man's use of fossil fuels as a source of energy.

More recently, the phrase climate change has been used as a catch-all term for the recent changes in overall weather phenomena seen around the world, largely associated with increases in global warming. The terms climate change and global warming are often used interchangeably. This is not strictly correct, but given the common use of both terms throughout these chapters, we have continued to use both terms in what is, for the most part, a harmless error. In fact, in the context in which we're addressing the issues, GLOBAL WARMING occurs for all the reasons that we've discussed, and this has led to CLIMATE CHANGE.

DEFINING GLOBAL TEMPERATURE

Finally, it is important to define what we mean by atmospheric temperatures and especially average global temperatures. Where is the temperature being measured? Is it surface temperature? Is it the air temperature just above the surface? Or is it the temperature at some distance in the atmosphere? Throughout this book, we have stuck to the internationally accepted norm where the term global temperature refers to the Earth's average air temperature measured at a standard height of 1.2 meters or 4.0 feet above the ground surface. Obviously, this number is going to change markedly depending on what season you're in, what the weather conditions are, what your elevation is, etc. But remember, here we're talking about averages, and the average global temperature data that we're using comes from actual measurements made at any given time taken at thousands of observation stations around the world, including deserts, mountain tops, urban centers, tropical areas, oceans, the Arctic, and the Antarctic, etc., all taken at specific times.

IT'S NOT JUST CARBON DIOXIDE

Throughout this book, we're focussing on carbon dioxide (CO_2) as perhaps the major greenhouse gas (a substance that "traps" the heat from the sun in our atmosphere). It is of course associated with the burning of fossil fuels and global warming, but it's important to note that it is not the sole culprit. Methane

(CH_4) and nitrous oxide (N_2O) are also important greenhouse gases that are emitted into the atmosphere largely as a result of human activity (e.g., agriculture, burning fossil fuels, industrial waste management), but also from decaying organic or vegetative matter, including of course trees. Water vapor represents yet another important and abundant greenhouse gas. There are many sources of water vapor in the atmosphere, and there has been some water vapor in the atmosphere since "the beginning." But there are two "new" significant sources of atmospheric water vapor that are occurring largely since detectable global warming really started some 50-plus years ago. One is that produced by humans through industrialization, and the second, interestingly enough, is caused by climate change and global warming itself in what is a disconcerting positive feedback loop. Think about it for a moment: As the temperature of the atmosphere increases, more water evaporates from our oceans, rivers, lakes, reservoirs, and any other sources of open water. This is an unfortunate positive feedback loop because as more global warming occurs, more water vapor enters the atmosphere, which, acting as a greenhouse gas, traps more of the sun's radiant energy in our atmosphere and produces even more global warming.

So, let's now look at climate, including both atmospheric CO_2 and global temperatures as a function of time, and let's go back even as far as the "dawn of creation" or the Big Bang, or whatever your religious leanings are, some four to five billion years to get a better understanding and then travel back to the present.

*   The temperature reference point as opposed to actual temperatures is important. It was established by the IPCC (International Panel on Climate Change) adopting a baseline reference period of 1850-1900 as 0*C. The numbers above that for example +0.5 refers to 0.5 degrees Celsius above the zero baseline (i.e. warmer) or -0.5, referring to 0.5 degrees Celsius below the baseline (i.e. colder). Note that we've been using the international measure of temperature in degrees Celsius or *C. If you insist on using the Fahrenheit scale, you need only multiply by 1.8 to convert from degrees Celsius to degrees Fahrenheit. So, an increase of 1.0*C is the equivalent of an increase of 1.8*F.

	Atmospheric CO2 (ppm or parts per million)	Global Temperature (*C relative to 1900)
4,000,000,000 BCE	40,000*	+20*C

These are of course just estimates drawn at a time when the sun was much dimmer (about 70 percent of what it is today), offering up much less UV radiation and much less "heating" in spite of the levels of CO_2 in the atmosphere. There is fossil evidence that cyanobacteria flourished in this era. These are the first "photosynthetic" bacteria, which utilized CO_2 and sunlight as their source of energy and generated tiny amounts of O_2, which eventually, three billion years later, would lead to life on Earth as we now know it.

500,000,000 BCE	4,000	+20*C

It's still hot, but things are about to change. This was a period of a rapid increase in biodiversity on Earth during a period called the Cambrian explosion. This is associated with an increase in oxygen in both the atmosphere and the oceans, increasing from near 0 percent to roughly 5 percent. The term "explosion" is a bit of a stretch since it occurred over a period of tens of millions of years, but the concept is correct…in time, there is enough oxygen in the air and in the waters to support life forms and evolution.

3,000,000 BCE	400	+2.7*C

By now, the Earth's atmosphere, especially CO_2 levels, have changed to an extent that it is not that far off from what we experience today. The sun's brightness is pretty close to what exists today. The Earth has begun to resemble its present state with all forms of animals inhabiting the air, the waters, and the land. There are even some early forms of four legged animals.

400,000 BCE	275	+2.0*C

The periods between 400,000 BCE and today represents the glacial and interglacial periods (the coming and going of ice ages) with lower atmospheric

CO_2 levels and lower temperatures during periods of intense glaciation and CO_2 levels increasing, but just modestly during periods of warming. The ice ages, each lasting at least 50,000 years, are caused by periodic (very long term) changes in the Earth's tilt and orbit, resulting in areas of the Earth getting less or more solar radiation and therefore causing the Earth to enter into a cooling (glacial) period or a warming (interglacial) period. The difference in response to these changes in the northern and southern hemispheres reflects the different proportions of land and ocean in the two hemispheres.

350,000 BCE   180   -5.0*C

Note below the 50,000-year cycling of the "ice ages" as the CO_2 levels moved between approximately 250 ppm, the warm periods when the temperatures were roughly +6 degrees above the norm and the glacial periods when the CO_2 levels were just 180 ppm and the temperatures were roughly -2 degrees below the norm. Many species of plants and animals became extinct during this period of extreme cold as ice covered much of the northern hemisphere.

By 300,000 BCE, Modern Man has begun to make his presence known in different areas of Africa, Asia, and Europe.

Year	CO2 (ppm)	Temp.
300,000 BCE	250	+2*C
250,000 BCE	180	-6.0*C
200,000 BCE	250	+2.0*C
150,000 BCE	190	-6.0*C
100,000 BCE	275	+3.0*C
20,000 BCE	185	-7.0*C
10,000 BCE	260	-1.0*C

The cycling of atmospheric CO_2 levels and global temperatures during this period represent the coming and going of the glacial periods, the last of which peaked around 20,000 BCE.

Between 10,000 BCE through to as recently as 1900 CE, the atmospheric CO_2 levels and the average atmospheric temperatures remained virtually constant, and that would have likely remained through to today and probably well beyond were it not for the fact that the Earth's human population increased from somewhere between one and 10 million (in 10,000 BCE) to 1.6 billion in 1900, when the Industrial Revolution began to take hold.

	Atmospheric CO2 (in ppm)	Global Temperature	Earth's Population
5000 BCE	240	-0.40*C	10,000,000
0	240	-0.20*C	170,000,000
1000 CE	240	-0.25*C	275,000,000
1600 CE	240	-0.35*C	500,000,000
1800 CE	265	-0.40*C	1,000,000,000
1850 CE	275	-0.20*C	1,200,000,000
1900 CE	280	-0.20*C	1,600,000,000
1950 CE	300	0*C	2,600,000,000

While it's not likely that many took any notice at the time, but by 1950, not only had the population increased to 2.6 billion, but the CO_2 levels had also inched up to 300 ppm, levels that had not been seen in millions of years and certainly not since the beginning of human life on Earth. While it was yet to be appreciated in 1950, it would become very evident that this correlated very tightly not only with the growth of the human population, but more importantly with the growth of the Industrial Revolution and the burning of fossil fuels to energize that revolution. In the early 1900s, it was coal, but by 1950, the use of oil and gas were becoming the essential sources of energy.

THE CRISES IS DEFINED

1950 CE	300	0*C	2,600,000,000
1975 CE	325	+0.15*C	4,100,000,000
2000 CE	365	+0.45*C	6,100,000,000
2022 CE	419	+1.10*C	7,900,000,000

By 2022, the world population had increased to 7.9 billion, the atmospheric CO_2 stood at 419 ppm. Global warming is a clear fact, and many believe that we have a full-blown climate crisis and that humanity is in trouble. These carbon levels are not only the highest levels in human history, but higher than at any time over the past three million years. Similarly, the average global temperature is higher than it's been in 200,000 years.

Projections are not easy. But if we carry on the course that we're currently headed (i.e., no mitigation of fossil fuel usage and continued increases in atmospheric CO_2), the numbers might look as follows:

2050 CE   480   +2.78*C   9,800,000,000

And many would argue that this is not compatible with sustainable human life on Earth, and we'll demonstrate some of these scenarios in the next chapter.

But there is still within our populations a serious problem of perception. Most of us, at least those of us living far from the equator, often experience significant extremes of local temperatures. Huge differences occur between cold winter and hot summer days. We can even experience changes in temperature 15-20*C and even more on any given day in any season. So how can increases in global temperature of even 1 or 2*C represent an existential threat to humanity?

We start by remembering the 10 items of change listed earlier in this chapter that are occurring even as the Earth's temperature has increased to date by just over 1*C. It's clear that weather-induced disasters are increasing yearly. How much more personal, financial, and physical damage can we sustain even at current rates if we're living in the fire-prone regions in the western United States, or in the hurricane and flood-prone regions of the southeast? We must think about how much worse it can actually become, while at the same time figure out how to stop and reverse climate change by ending our continued use of fossil fuels as a source of energy.

GLOBAL WARMING AND EXTREME WEATHER
In examining the "crises" list presented at the start of this chapter, most of the items would not strike those of us who live in quite temperate climates as obvious on a day-to-day or minute-by-minutes basis. They accrue over time and then they sort of sneak up on us, and then it hits us – "Wow! That's really serious!" when we see it on TV, or if we lose our homes, our livelihood, or worse of all, a family member. We ought to be paying greater attention.

While the relationship between population growth, industrialization, the burning of fossil fuels, the increases in greenhouse gases and the resulting global warming may be straightforward, the relationship with extreme weather is anything but.

While we are individually unable to detect, the relatively small changes in atmospheric temperatures, or the small rises in ocean levels, the dramatic increase in extreme weather events are and will continue to be very evident.

It is of course too easy and not very productive to simply relate to anecdotes or individual events as in:

> That fire in California was the worst ever.
> The flooding in Louisiana was devastating, as was the cost to rebuild.
> It's never been so hot; the temperature high beat all previous records.
> It seemed that in January 2020, most of Australia was on fire.
> When I was a kid, the snow was sooooooo high!
> The polar vortex made this week in the Midwest the coldest ever.

That's all well and good, but there is a saying that goes something like: **"The plural of anecdotes is not data,"** and if we are to take the issues of global warming and extreme weather seriously and take dramatic and likely expensive steps to counter, then we're going to need real and convincing data.

WHAT IS THE EVIDENCE FOR A GLOBAL INCREASE IN EXTREME WEATHER?

First, let's establish that we're dealing with facts, not opinions. A weather event is defined by official weather authorities to be- extreme, if it is unlike 95 percent of similar events that have occurred in roughly the same area. Examples of extreme weather events might include drought, extreme heat, extreme precipitation, hurricanes, tornadoes, floods, wildfires, etc., but only if they reach that 95 percent threshold.

That the number of extreme weather events has increased dramatically over the past 70 years is clear. Using the year 1950 as the base, extreme weather events increased by the following:

Increases in Extreme Weather Events

1960-1970	7%
1970-1980	11%
1980-1990	18%
1990-2000	36%
2000-2010	41%
2010-2019	62%

These are impressive, convincing, and alarming numbers. But they're not just numbers; each of them has had devastating social and economic ramifications. There has been a dramatic increase in heat waves across the world from 1980 to date. In Texas and in areas of the Midwest, the number of extreme heat days has more than tripled (that is an increase of 300 percent) between 1980 and today. And these numbers are reflected worldwide. In Europe, the number of severe floods in 2018 was the highest ever recorded. Extreme weather events in Africa are becoming more frequent: the Horn of Africa is experiencing long period of extreme drought, serious storm-inducing floods in East Africa, which are having devastating affects on agriculture. There have even been snowstorms in the high altitude areas of the Sahara Desert, resulting in flooding and erosion.

China, with its huge landmass, is experiencing dramatic increases in extreme weather events with devastating impacts on its social infrastructure and economy. At the same time, huge increases in rainfall and destruction of agricultural areas are being experienced in the southwest province of Yunnan, while at the same time large areas of central and northern China have been devastated by drought. The lowering of water tables as a result of drought has resulted in huge areas in central China becoming agriculturally infertile and largely evacuated. The loss of water has not only crippled hydroelectric power plants, but it has also resulted in repeated crop failures over the past dozen years. To be accurate, not all this devastation can be attributed solely to global warming. As an example, over a period of not so many years and massive population growth, the Chinese, like many others around the world, have been very poor stewards of their forests, and deforestation has proven to be a major factor in the preservation of ground water. In turn, the resulting arid areas can have an important impact on weather and weather-related events.

FIG 19 ONE OF THE MANY FACES OF EXTREME WEATHER

The relationship between cause and effect is often not straightforward when it comes to climate change. Flooding has become more prevalent worldwide, even in areas where the total annual rainfall has actually decreased as incidents of flash floods, urban flooding, river flooding, and coastal flooding are on the rise around the world. The costs to society can be very significant as damage occurs to various structures with enormous clean-up costs. Coastal and river flooding can cause enormous damage to agricultural areas, including when saltwater flooding contaminates fertile growing areas, rendering them virtually barren for further planting.

Between 1980 and 2009, worldwide flooding alone caused more than 500,000 deaths, affected 2.8 billion people, and in the United States alone resulted in property and crop damage averaging in excess of $10 billion per year. That number reflects only damage from floods and does not include other extreme weather costs (e.g. fires in California, hurricanes in Puerto Rico and Florida, etc.)

From the early 1980s until 2019, we saw a significant increase in the number of hurricanes as well as in the intensity of hurricanes, with many more Category 4 and Category 5 storms being reported in the more recent years.

It may seem counterintuitive (this is covered in the next section) in the context of global warming, but the incidents and intensity of severe winter storms has also increased. There have also been substantial increases (from the early 1980s to the late 2010s) in the frequency of other severe storms, such as tornadoes, hail, damaging thunderstorms, and wind. The patterns of these storms have also changed substantially, as areas that rarely encounter severe weather, thunderstorms, and rains now see them routinely. The notion of a polar vortex bringing freezing weather to areas of the southern and midwestern United States was a rarity not many years ago whereas today it appears once, twice, or even more often each winter and with increasing degrees of severity.

WHAT IS THE RELATIONSHIP BETWEEN GLOBAL WARMING AND EXTREME WEATHER EVENTS?

It's easy to document the incidence, severity, and impact of extreme weather events, but really understanding the link of those events to global warming is much more difficult. To start with, it's clear that the warming of the atmosphere can trigger a whole series of both simple and complex weather and climate events. As an example, on the simple side, global warming results in increasing the water vapor in the atmosphere (the water evaporation coming from lakes, rivers, and oceans, as well as the land), which results in more frequent and heavier rainfalls and snowstorms. When the atmosphere is warmer and contains more moisture over the oceans, hurricanes that develop tend to be more intense, larger, and produce more rainfall. Global warming also causes sea levels to rise, which results in more devastation to shorelines (cities, town, and agricultural land) during coastal storms. When the warmer air increases evaporation and this occurs over land, moisture, may be drawn from the land producing drought conditions. This in turn can lead to further erosion of the soil, destruction of agricultural lands, loss of animal habitats, and the deterioration continues. These are no longer isolated events.

These are the straightforward impacts of warmer air and the impact of more moisture in the atmosphere. But weather is much more complicated than that, as is the global impact of global warming. As an example, parts of the United States in recent years have experienced record cold temperatures and record snowfall, all of which have been blamed on a "polar vortex" sitting for a week or more over states that have seldom seen such cold weather or snowfalls and are usually ill-prepared. How can that be caused by global warming? A commonly accepted explanation is as follows:

It has been determined that the Arctic is warming approximately twice as fast and sometimes even more than other parts of the Earth's surface. This has been referred to as Arctic amplification, and is thought to be due to the rapid loss of sea ice coverage in the area. The warmer air in the Arctic is thought to push cold air masses southward, resulting in warmer weather in the north and colder weather in the south in the middle latitudes. The evidence for this explanation is mounting year by year as greater amounts of the Arctic become ice-free during the northern summers. There have been days in recent years where temperatures recorded in towns situated above the Arctic circle have been warmer than those at the Canadian-American border (e.g. Vancouver, Seattle, and Chicago).

Another theory suggests that the position of the jet stream, a band of rapidly flowing air high up in the atmosphere, is driven by the temperature difference between cold air in the north and the warmer air in the south. With global warming and the warming of the atmosphere in the north, if the jet stream shifts more to the south, then it will pull more of the cold Arctic air south, and this may explain the polar vortex, cold weather, and snow experienced further south than ever before. As is often the case, the two explanations may be complimentary. This is important to understand, because how else would someone from Texas or Missouri have any understanding that the cold, frigid weather and bursting water pipes they are experiencing in February possibly be attributed to global warming?

Clearly, these are not simple issues. But what seems sure is that global warming doesn't only mean that our atmosphere is getting warmer, but it also has a major impact on weather patterns, and the weather becomes "weird." That is, we experience more extreme weather events, most often not to our liking. It is still important not to confuse "short-term weather events" in selected areas of the world with longer term changes in the climate and overall global warming.

We are left with clear evidence that global warming is real, that climates have changed, that weather patterns have changed, and that extreme weather events have become more common.

THE EL NIÑO

It is also important that we not revert to "blaming" global warming every time there is a weather event. Storms and strange weather events have always occurred, and until recently, they had nothing to do with the changes in our climate resulting from our dependence on burning fossil fuels.

The El Niño weather phenomenon is an excellent example. It was originally described as an event seen and experienced by fishermen off the coast of South America and reported as long ago as the early 1600s. They described the periodic appearance of unusually warm water in the Pacific sometime in December or around Christmas. "El Niño" is Spanish for "the boy child," in reference to the birth of Baby Jesus during this period of the year. This is perhaps the best example of the complex interaction between the ocean and the atmosphere. It goes something like this:

Periodically, every five years or so, warm ocean currents in the Pacific (west to east) with temperatures as much as 6*C above normal cause warm water to build up around the equator off the western coast of South America. This warms the atmosphere, and massive amounts of water vapor from the oceans move into the atmosphere, resulting in rainy weather and tropical, sometimes severe, thunderstorms moving up the west coast to many parts of Central America, Mexico, California, and even further north. Devastating storms, loss of life, and even the introduction of vector-borne diseases such as malaria are sometimes seen. At the same time, areas of Southeast Asia and Australia, where the warm water originated, exhibit just the opposite weather system as colder ocean water temperatures result in drier air, severe droughts, and terrible wildfires.

The opposite of the El Niño is the La Niña, also cyclical in nature and dependent on ocean currents where the atmosphere cools in response to the cold pacific waters along the equatorial coast. Under these conditions, there is less water in the atmosphere, and rainfall dramatically falls in areas such as Ecuador, Peru, Mexico, and the southwestern United States. La Niñas are often associated with drought conditions in those areas and corresponding terrible flooding in parts of Southeast Asia.

It is clear that the severe weather patterns brought on by the El Niño are independent of manmade global warming. What is less clear is whether or not global warming intensifies the storms brought about by El Niño. Some feel that in recent years, El Niños have occurred with greater frequency, every two to three years as opposed to the historic four-to-five-year cycle. But the jury is still out, which means that we simply don't have enough data.

In thinking about climate change, one also might think about Mother Nature and her (is that still PC?) role in all of this. It would certainly weaken our case for the need to change human behavior in the current climate crisis, if we were to simply ignore examples of climate change that are clearly not manmade. There have been other climate change events that have occurred over the past half century or so that might clearly be placed at the feet of Mother Nature and not be blamed on our penchant for burning fossil fuels and causing global warming. The El Niño effect just described is one of them. The dust bowls that occurred across the southern Great Plains of North America in the 1930s are an interesting case in point. The dust bowl represented a period when the regions suffered from extreme wind erosion, basically laying bare more than a hundred million acres of farmland. It was a period of low crop prices (it was in the midst of the Great Depression) when the farmers didn't have the wherewithal, or the knowledge, to implement soil preservation programs on their lands. Severe droughts and extreme winds produced devastating events on the agricultural productivity of huge areas of the United States and some in Canada. Many of the areas, in fact, never recovered.

While there is not a lot of data, reports out of China suggest that similar manmade arid conditions have been produced on the central plains. These events may have been manmade but cannot be blamed on the burning of fossil fuels.

Volcanoes have also been known to affect climate change. Huge amounts of volcanic gas and ash are injected into the stratosphere. Most of the ash falls rapidly to Earth with little effect on the climate. But the various gases can have multiple effects. Atmospheric SO_2 (sulphur dioxide) can cause global cooling by reflecting sunlight back into space, and when combined with water (H_2O), it can produce sulphuric acid (H_2SO_4), resulting in acid rain. Volcanic CO_2 can act as a greenhouse gas and result in global warming. Generally, this eruption is not large enough to have any permanent or global impacts. The eruption from Mount Pinatubo in the Philippines in 1991 injected 20 million tons of sulphur dioxide into the stratosphere and cooled the Earth by 1.3 degrees Fahrenheit or 0.5 degrees Celsius for about three years after the eruption. This heating was temporary and local. It is important for us to understand (and acknowledge) that not all adverse weather events can be attributed to global warming resulting from our excessive dependency on fossil fuels.

Chapter 16
DEFINING THE CRISIS

The word "crisis" is used repeatedly when climate change is being discussed or debated. The question is, what is a real crisis, and what does it mean, and when is it just warmer weather? And are we being overly dramatic in using terms such as "existential crisis"? And when are we simply taking on a cause, like tree hugging or saving the whale and overstating the impact on humanity simply in support of one of our own favorite hobby horses? Many of us, especially those of us living relatively affluent western lives, will have trouble relating to the notion of the world being in a serious climate change induced crisis with our very existence being in jeopardy. If we are sitting in a Manhattan apartment about to go out to dinner and a play and there's a severe thunderstorm outside, it's hardly a crisis. If, on the other hand, we live in California and we've just been evacuated and see our neighborhood burning down in the rear-view mirror from a hillside fire (the second in three years), and insurance wasn't coming through , you better believe that that's a major crisis, at least on a personal level.

Greta Thunberg dramatically stated to the World Economic Forum in Davos in 2019 in her call to save the world from the ravages of manmade climate change and global warming that, **"Our house is on fire."** She was at the same time looked on by many as an incredible young hero and the voice of the next generation and simultaneously as a silly young alarmist. But increasingly, her voice has credence. The United Nation's Intergovernmental Panel on Climate Change (IPCC) in 2019 issued its most stark report and dire warning. It stated that: **"The climate crisis is reducing the land's ability to sustain humanity."** That certainly sounds like a crisis and should stimulate us to ask a number of simple questions:

1) Where do we stand today in respect to the crisis?
2) What is the "trajectory" of the crisis?
3) Is there a time when it will be too late to make corrections to avoid the crises?

WHERE DO WE STAND TODAY?
For 800,000 years, the amount of carbon dioxide in the atmosphere never rose above 300 parts per million (ppm). In 1950, it crossed that line, and by 2019, it had reached a level of 417 ppm. Since 1950, the planet's average temperature has risen by 1.5 degrees Fahrenheit or 0.85 degrees Celsius.

Now the scientific evidence for the warming of the climate system is unequivocal, and so is the fact that it is manmade caused by the extraction and burning of fossil fuels.

The repercussions of these changes are pretty clear: warming oceans, shrinking ice sheets, glacial retreats, decreased snow cover, sea level increase, declining Arctic sea ice, extreme weather events, ocean acidification, and the list goes on...

But do these phenomena really reflect a crisis situation of world proportion, or are these disruptions just "local" in nature? Can we really talk about impending doom, as in "our house is on fire," based on those changes? Again, it's probably hard to think about the crisis from the comfort of many of our airconditioned and stylishly-furnished homes.

But you do sense and experience that we are in a real crisis if:

- You're a refugee fleeing the drought and famine of Sub-Saharan Africa and floating in an overcrowded dingy in the Mediterranean, looking for a safe place to land.
- You're a rancher in Australia, and your whole ranch and all your animals have been destroyed by fire.
- You live on one of the Abaco Islands in the Bahamas, and Hurricane Dorian has destroyed your entire island.
- You live in a part of northern China, where it hasn't rained in years, and your entire livelihood as a farmer has been decimated.
- You're a farmer in Northern India, and heavy monsoon rains have flooded your fields, destroying your crops, and your fields are flooded by incoming sea water, leaving your fields infertile (due to the saltwater).
- You're a polar bear, and you can't find an ice flow from where to fish for harp and hooded seals.
- You're a coral reef, and you're being destroyed by the acidification of the oceans.
- You have to shutter your family's California winery because you can no longer get insurance.
- Your insurance rates triple because your insurance company has incurred so many losses.
- You've just lost your home and all your possessions in Dallas, Texas, to fire following a polar vortex-induced deep freeze, and you can no longer afford to be insured.

These are all very real crises, and the data shows clearly that the incidents of these events is increasing sharply. You might also think about the costs that insurance companies, financial institutions, and governments are incurring as a result of the severe weather events and ask how these costs can be sustained. But then again, unless they're harbingers of things to come, you might have a hard time seeing them as somehow being existential or an actual threat to the real existence of human life on Earth.

WHAT IS THE TRAJECTORY OF GLOBAL WARMING?

At the current rate of increase of atmospheric CO_2 (3 ppm/year), by 2050, the atmospheric CO_2 will have increased to 500 ppm+ and the global temperature will have increased another 4.0*C. That's really bad news, even suggesting the doomsday scenario put forward by some, as you'll see below.

But let's for a moment take a smaller step. The current temperature increase (global) from the 1950 baseline is +1.01*C. And that number or that degree of global warming has already caused serious damage to many parts of the world.

What would happen if the number became 0.7*C warmer or up to 1.7*C above the established 1900s level? Just before the start of this list, it's important to remember that we're talking about averages for the whole Earth. The numbers may vary dramatically. In the Arctic, the number might be as high as +5.0*C whereas over the oceans it might be just 0.5*C.

Unfortunately, notwithstanding our grave concerns and some of our mitigating efforts, it is likely that we will move from the current +1.01*C to +1.5*C warming within the next 10 years, and the impacts will be substantial, if not critical, for many. Many estimate that these scenarios are likely even if we begin efforts to reduce fossil fuel emissions immediately. Largely because it will take some time to "turn off" the progress of global warming once we've made the commitment to do so in a serious manner.

Severe heat waves will be experienced by many more people on Earth with the number of people being exposed increasing from 14 percent to 37 percent over the course of the next 10 years. These episodes of extreme heat will result in millions of deaths, especially amongst the poor in developing and underdeveloped nations where there is little, if any, escape from the heat. And for those of us living in relative affluence, we have only to sit back and reflect for a moment on the number of extremely hot summer days we've been experiencing in recent years. Or if you live above the Arctic Circle in Siberia and the temperature hits a high of 38*C or 100.4*F, as it did in June of 2020, that number alone should raise alarms in all of us.

Water will become extremely scarce for at least another 500 million people across the globe. Severe droughts and the destruction of agricultural lands, in a manner currently seen in major parts of northern Africa, central Asia, Australia, and areas of South America, will increase even more.

Extreme precipitation is the flipside of drought, and many areas in North America and areas in northern Europe and northern Asia will experience very high and continuous periods of rainfall resulting in serious mountain flooding. If this isn't a form of extreme weather, it is at least a major disruption of normal weather patterns causing immense damage.

Ocean levels will increase by an additional 1.0 meter, resulting in coastal flooding, beach erosion, salinization of water supplies, and the destruction of huge amounts of agricultural land. Major coastal cities like New York, London, San Francisco, Singapore, Venice, and many others will be exposed to increased and extended flooding. Costs will be horrendous. The financial threats to the bottom lines of insurance companies and financial institutions, and by extension to all of us, have already begun to occur in earnest.

Oceans will become even more acidic and contain even less oxygen, resulting in large parts of the oceans becoming effectively dead. This is already a major problem. Ocean fisheries will be reduced by 75 percent and more. It's indicative that while in 1970, only 5 percent of all the fish we ate came from fish farms, in 2020, that number worldwide was well over 60 percent. It is difficult to know how much of that change came as a result of overfishing and how much can be attributed to the loss of marine life as a result of global warming and ocean acidification.

Biodiversity…we have all heard how climate change is destroying whole ecosystems as thousands of animal and plant species become extinct largely due to loss of natural habitat. The rate of this extinction goes up by an additional 18 percent as global warming approaches +2*C.

Humans will be massively impacted. Heat-related illness and mortality will go up sharply. Vector-borne diseases, such as malaria and dengue fever, will increase dramatically as the insect carriers move to populated areas north and south of the equator where the population has never been exposed to many of these "foreign" viruses and other pathogens and therefore have little resistance. Food security and famine will be ever more important issues in major areas such as Sub-Saharan Africa, Southeast Asia, and Central and South America. The attempted migration of political, economic, and climate refugees would increase sharply, creating additional civil unrest in more fortunate and/or affluent parts of the world. Whole economies would be destroyed, and the GDP of wealthy western countries in Europe and the Americas would suffer major drops.

Obviously, these are dreadful scenarios. But they would pale in comparison to what would happen if the temperature increased by a whole +5.0*C. That would be a real doomsday scenario for vast populations. We'll describe that only briefly later in this chapter.

IS THERE A TIME WHEN IT'LL BE TOO LATE TO MAKE CORRECTIONS?

Yogi Berra, of baseball fame famously said something to the effect that:

"Predictions are very difficult to make, especially about the future."

This rings only partially true. When it comes to climate change, there are so many variables, many that we are aware of, some that we aren't even aware of, and most of which we are yet to have any control over. But there are some things that we do know for certain:

A) Climate change and global warming are real and manmade.
B) Climate change has wreaked havoc with hundreds of millions of lives and caused untold damage to our physical environments.

C) Most notably and quite incredibly, we have collectively done very little to combat climate change.

With all of that in mind, there is every reason to believe that without introducing changes, the pace of global warming will increase, and we should only expect that things will get worse.

Now as bad as things are, and they are bad, there's still reason to believe—or at least hope—that we have not yet reached any major tipping points when we will have lost the possibility of reversing the impact of climate change. That would be the point where we could no longer control the increased rate of global warming, and where we had no control over the amount of greenhouse gases being poured into the atmosphere. Those are obviously "what ifs," but for now, things are bad and only getting worse.

THE CRISES OF FEEDBACK LOOPS AND TIPPING POINTS

Let us briefly explain those two ideas. The feedback loop, as in a positive feedback loop, would be like getting too close to a microphone. The closer you get, the louder your voice appears, and the more interference there is. The tipping point simply reflects a point of no return: You do something, and there is no going back. Say you go through a door; you lock it and throw away the key. That's a tipping point. You made the decision to go through that door, and there's no way to reverse the decision. A positive feedback loop with an approaching tipping point is a frightening thing, and in terms of climate change, it might be fatal to ourselves or to some part of our Earth.

Another analogy is that of the runaway train. The brakes have ceased to work; it's all downhill, and the train's speed is increasing. A wreck is coming and can't be prevented. It's truly the doomsday scenario.

Here are a few examples of feedback loops and tipping points associated with global warming, CO_2 levels, and global temperatures.

Water Vapor: This is the simplest and the most common of the feedback loops, and it goes something like this. Global warming causes water to evaporate from the Earth's vast expanses of water (oceans, lakes, and rivers). The water vapor functions as a greenhouse gas, causing more of the sun's heat to be trapped within the atmosphere. This causes the air above the bodies of water to warm even more, causing more water to evaporate, producing more greenhouse gas (the water vapor), causing even more heating and further temperature increases, hence the positive feedback loop.

The feedback loop could of course work in reverse, but it would require an actual cooling. That might occur if we could figure out a way to actually decrease greenhouse gases, primarily carbon dioxide, methane, and water vapor.

Methane and the Arctic Permafrost: This may, in fact, be the most concerning and disconcerting of the positive feedback loops, mostly because the timing and the size of the problem are not well defined. That does not make it any the less worrisome. The loop starts with global warming, resulting in a dramatic decrease in the ice/snow coverage in the Arctic. The resultant exposure of the permafrost causes a release of carbon in the form of methane (CH_4) into the atmosphere. CH_4 is an even more potent greenhouse gas than CO_2, and so the atmosphere warms even more, resulting in more melting of the icepack, more exposure of the permafrost, more release of methane, more greenhouses gases, and higher atmospheric temperatures…and the positive feedback loop goes on. That the permafrost has the potential to release large amounts of carbon in the form of methane is not surprising since these areas are replete with peatbogs, decayed and decaying vegetation, and other organic matter that lay frozen under the ice and snow for millions of years.

Recent reports (2020) suggest that the extent of warming in Siberia and the exposure of the permafrost and the release of methane may be happening faster than we thought. And that's a major problem if we aren't successful in stemming the tide of global warming.

Ice Melting: When ice melts, as in the case of the far north and far south, land and water take the place of the ice. The land and the water are both less reflective than the white ice, and thus absorb more of the solar radiation. This, of course, results in even more warming. More of the ice melts, less of the solar radiation is reflected back into space, more solar energy is absorbed, causing additional warming, and the cycle continues. Global cooling would reverse this, as additional ice coverage would increase the amount of solar radiation reflected back into space, decreasing the amount of solar radiation absorbed and thereby resulting in global cooling.

There are a multitude of such feedback loops that are occurring as a result of global warming, none of which represent good news. In a word, they should frighten us. We should be worried, and we should have a clear call to action.

THE DOOMSDAY SCENARIO (IN BRIEF)
We would be remiss if we didn't at least briefly describe the doomsday scenario. What would happen if the global temperature rose by +8*C? Suffice to say, it wouldn't be pretty. Some have called it a climate apocalypse; a catastrophic event, or series of events in this case, leading to the collapse of human civilization. It might even be thought of as a final or dystopian version of Moses' 10 plagues.

Atmosphere: Warming oceans and the complete loss of ice coverage in the northern and southern poles, for greater parts of the year, result in the release of massive amounts of the highly flammable methane that can be ignited by a simple spark or lightning, resulting in massive and even apocalyptic fires.

Disease: Increased temperatures, food scarcity, and altered vector ecology lead to major epidemics of diseases that we have little or no defense against.

Heat Deaths: Most plants and animals cannot sustain life consistently above 37*C. Temperatures rise routinely above these levels in increasingly larger parts of the planet, and death by hyperthermia (overheating) becomes prevalent.

Water Scarcity: Huge areas of the planet are without any forms of fresh water, making them increasingly uninhabitable to the vast majority of plants and animals, including of course humans.

Starvation: Between population growth and physical disasters, we are simply unable to grow/produce enough food to sustain ourselves, and the number of people dying of starvation increases dramatically.

Flooding: If all the ice on land and at the poles were to melt, our oceans would rise by 65 meters and over three billion people would be displaced, to say nothing of the complete devastation of agriculture associated with worldwide coastal flooding. Many coastal or sea level cities would effectively disappear (e.g., New York, Los Angeles, Miami, London, Paris, New Delhi, Rio de Janeiro, etc.).

Displacement of Societies: Massive worldwide migration occurs as a result of the huge change in resources available to sustain life.

War: In the early days of this apocalypse, nations and people go to war against each other in order to claim or retain resources. Water becomes perhaps the most precious commodity to be fought over.

Collapse of Societies: As the means to sustain life is challenged, wholesale changes in societies across the globe occur as groups move to seek required resources.

Mass Extinction: Nothing more needs to be said. There will of course be microbes and perhaps some species of plants that will survive, but little else. And then we can only surmise what will happen to life on Earth in the millennia and millions and billions of years that follow. It would probably recover, a cooling period might even occur, but it certainly wouldn't be the earth that we know.

Interestingly, there is a global environmental movement called Extinction Rebellion. They are a nonviolent group who use very aggressive tactics of civil disobedience to compel government action to avoid the tipping points in the climate system to avoid the extreme loss of biodiversity and the very real risk of ecological and social collapse. There is no doubt that the group is extreme. You might even question their tactics, but it's very hard to argue with many of their statements about what might occur if we don't take massive action and soon. The following is their manifesto:

> *Under our current system, we are heading for disaster. Catastrophic climate change will kill millions, cause food collapse, and render more homeless. Massive extinction of wild species will lead to ecological collapse. Destruction of natural habitats will lead to genocide of indigenous peoples and the loss of our planet's support systems. It's not too late to change course – a better future is possible. But governments are consistently failing to take the urgent decisive action that will save us. If the system will not change, then we must change the system.*

Now if that's not a bit depressing, Greta Thunberg helped to drive it home in what she famously said in her United Nations Speech: **"I WANT YOU TO PANIC."**

Panic, but don't despair. There are solutions. If we, and especially our leadership, put our minds, our resources, and our science to it.

Chapter 17
THERE ARE SOLUTIONS

It's not lost on us that all that was more than a bit depressing. And then when you hear the loud voices of the climate change deniers and some of the political and corporate strength that they represent, you almost want to stick your head in the sand and ignore global warming. Well, obviously that sort of fatalistic approach is NOT a good idea. Humanity really is in crisis, and we have no choice but to deal with it.

No apologies will be made here for painting such a dire picture of the future if we don't take the necessary corrective steps. We can't predict with any accuracy when or if any one of the various doomsday scenarios will occur, but if we maintain the status quo, with reasonable certainty, they will. Furthermore, there is no reasonable expectation to think there is a some sort of natural cure; for instance that the Earth will all of a sudden perform a tilt, allowing less heating by the sun, or that new trees might sprout up to suck untold amounts of carbon dioxide out of the atmosphere, decrease greenhouse gases, and start cooling the planet. That's just not going to happen. There is no reason to believe that the Earth, or Mother Earth to be more specific, can figure out how to heal herself, certainly not in the time frame that we have available.

We're going to have to create the solutions to reverse climate change and global warming, and there's every indication that we're going to have to do it fast. There are simply too many tipping points that threaten to bring the many different crises to a head, and which lead to irreversible events in the Earth's climate. Some of these are not consistent with the continuity of human life on Earth, and while that's a tough statement, we believe it to be true.

Before starting to address specific solutions to the climate change crisis, there are a number of generalities that have to be addressed. Huge segments of the world's populations, their governments, and their industrial leaders have to get on the same page. This simply cannot be a partisan or a "limited engagement" initiative. It must be global. Climate change and global warming issues do not bide my national boundaries.

Here is an example of one country that was going in exactly the opposite direction. Brazil is a country of 201 million people (the sixth largest in the world) with a land mass of 8.5 million square kilometers (the fifth largest in the world) and home to the largest part of the Amazon rainforest. Their current leader is a right-wing fascist Jair Bolsonaro, who is sometimes dubbed the "Trump of the Tropics," for his violent talk and far right-wing views on democracy, human rights, women, race, immigration, homosexuality, climate change, and the coronavirus pandemic. Here are a few of his quotes:

"I defend torture."

"The situation of the country would be better today if the dictatorship had killed more people."

Bolsonaro claimed repeatedly that the coronavirus pandemic was a "fantasy"; that it is nothing more than "a little flu" and encouraged pro-government and pro-church gatherings to fight against the notions of isolation and social distancing. On the issues of global warming and climate change, Bolsonaro addressed the United Nations in September of 2019, asserting that the forests in Brazil were "practically untouched," and he blamed a "lying and sensationalist media" for propagating fake news about their destruction. Now that sounds familiar, doesn't it? We are discussing the issue of Bolsonaro and his "out of this world views" to emphasize the degree to which solutions must be global. Too often in recent years, political and extremist views have replaced science, let alone common sense. And if the truth be told, Donald Trump and his many decrees relative to regulations, environment, and climate change were not any better. Simply put, the solutions must be international. They must encompass the vast majority of the Earth's people and the vast majority of the Earth's landmass. Finally, they will have to entail the doing away with the vast majority if, in fact, not all the current fossil fuel usage.

So then, what are the most important general requirements to succeed in this war against climate change or this battle to save humanity and to affect a number of solutions?

1) Education: When we talk about global warming and climate change, we have to constantly remind ourselves that this is a global problem. You have to consider seriously the phrase: "Think global, act local," because make no mistake about it, change has to occur, and solutions have to be found for the entire world. We have to get the WORLD on the same page. HUMANITY is in CRISIS; CLIMATE CHANGE and GLOBAL WARMING are the CAUSE, they are MANMADE, and they do represent an EXISTENTIAL THREAT. The data shown in earlier chapters is very clear. The Earth does not have the capacity to correct the insults in a timeframe that humanity requires, and so we have to find solutions that are also manmade. This is no longer a choice. This is an imperative. We have to take a lesson from Greta Thunberg and her Fridays for Future program to tell us how serious this all is. It may be that young people, who have more at stake, have to take to the streets to ensure their futures, as they have on many occasions around the world over the past half century. But we have to watch those demonstrations; we have to listen to what the young people are saying; and most of all, we have to act and in a timely fashion.

2) Cease and Desist: It seems pretty obvious that we have to do everything that we can to decrease—no, that's not correct: to STOP the emission of greenhouses gases through the burning of fossil fuels, and we have to decrease the continuing fouling of our planet. This is no longer simply an economic imperative. It is a human survival issue. We have potential solutions at hand. We just have to figure out a way to make the moves mandatory and/or penalize, as in tax, or even in the extreme, incarcerate, those who refuse to change and who continue to add to the problem. We also can't fall into the trap of thinking that we can simply use carbon taxes to subsidize clean energy sources until they become self-sustaining and to support research into new clean energy technologies. We cannot allow some to simply pay the price of continuing to burn fossil fuels and to continue warming the planet. That is SLOW THINKING, and it won't work. The incentives are simply not there, and we don't have the luxury of time.

The following is a set of examples of what must be done:

Institute an immediate halt to the burning of fossil fuels. Although that might be impossible today, surely it can be done in less than five years, and perhaps even two to three years if we have the political and industrial will. We simply can't afford to wait until, for instance, 2050. Where will our energy come from? That's straight forward: hydroelectric, solar, nuclear, wind, and hydrogen are all energy sources that are currently available. The immediate response of course will be that burning fossil fuels is much more economically viable… That's not an acceptable answer.

Not only is it not an answer, but it's not even true. The fossil fuel industry receives many billions of dollars of subsidized funding around the world. Stop those subsidies and make it economically penalising not to change. This is too critical an issue to even think about instituting the excuse not to change because it's too expensive. The facts are clear that the costs of not changing are much more severe. There will of course be significant start-up costs that our governments, our institutions, and our corporations must accept. And they must all understand that the cost of not converting our sources of energy away from burning fossil fuels is much, much higher.

3) Reforestation: We've dealt with the issue of trees and reforestation in an earlier chapter. We concluded that reforestation of the planet alone will not be a sufficient solution to global warming, but that is not to say that it isn't very important for a sustainable future. The planting of lots of trees, numbering in the trillions, not millions or billions would be an obvious improvement as the forests would effectively suck carbon dioxide out of the atmosphere. Suffice it to say that it would be a viable option in the long term, but it would require a massive change primarily in the way we grow food. Remember that deforestation had a particular economic reason. It's been "necessary" as the population grew to make room for new farms, highways, cities, and industrial parks. While planting trees will be critical, it's not reasonable to expect us to reforest all of those areas discussed above. For example, where would we grow our food. On a side note, there is actually an interesting development, a new technology called Cellular Agriculture which will be covered later in this chapter.

4) Research & Development: There are a whole host of solutions being worked on in our universities, our research institutes, and in a variety of different small start-up companies as well as large multi-national corporations, and these all have to be fast tracked. Many of these are much more than just "academic thoughts or ideas." They are solid solutions that have been looking for investors, and for those who sit in their corner offices, or perhaps through big money to understand their value. Ideas like reducing greenhouse gases by using CO_2 drawn from the atmosphere to manufacture cement and other essential building materials. Ideas like extracting nitrogen from the atmosphere to make nitrogen-based fertilizers rather than mining nitrogen-based

chemicals for use as fertilizers at great expense, especially energy expenses, and an expansion of the carbon-footprint. Many of these technologies have already shown great promise in pilot projects, but those that are worthy must be brought to the market rapidly; as in yesterday. In banning fossil fuels, we have to turn to well established alternative energy sources. The big five—hydroelectric, solar, nuclear, wind, and hydrogen—are all available and well understood in their own rights and will be covered later in this chapter.

But research has only begun to show and create new, exciting solutions. Here's a taste:

One of the most interesting areas follows a simple but quite astounding popular statement that, ***"In a single hour, the amount of power from the sun that strikes the Earth is more than the entire world consumes in a year."*** The question that follows is, why do we have to continue to burn fossil fuels and create greenhouse gases and global warming? Why can't we just capture a small amount of the sun's energy that is poured down on us every single minute? The reasons or answers to those questions are not very complicated. The most direct way that is currently available to capture that energy would be through the use of solar panels. That energy could be and is used to heat our water, to warm our houses, and to generate electricity to power our industries. Then why is it that only a small fraction of the energy we use on Earth is derived through solar energy and solar panels? The usual answer is that while current solar panels are pretty efficient at capturing the sun's energy (up to 24 percent of the energy can be captured), they are expensive, and furthermore, battery technologies that we have available to store that energy are poor. We don't have effective ways to store that energy for use at a later time, say at night or during those dull, cloudy days, when as they say, "the sun don't shine." In other words, we have not developed serious energy storage technologies as in more efficient, lighter, and higher capacity batteries. And why is that? Well, it's pretty simple. We've become very comfortable with the so-called cheap energy derived from oil, gas, and coal without concern for the long term consequences. Energy is cheap, and we have everything that we need. How very wrong was that.

In fact, very little effort has gone in to producing better solar panels and better battery storage systems. Simply put, the apparent need and the drive to do so just wasn't there. But it is now, and these efforts should pay incredible dividends if the investments were just made, and now. And as with most technology developments, with time, they get better, more efficient, and even cheaper.

There are also a number of exciting technologies that have passed the research and development stage and are in various phases of testing and scale-up. One such technology is being developed by the Dr. Craig Venter, who led the private sector arm of the Human Genome Project and is considered one of the fathers of the new discipline of synthetic biology. His company, Synthetic Genomics, and the mega corporation Exxon Mobil are developing a new source of oil (why is this not surprising?) that has the potential to closely approach a zero carbon footprint. That is, it doesn't result in a net increase in greenhouse gases. While this energy doesn't come from any of my (hi, it's me, tree again) now dead relatives (i.e., fossils), it does come from a cousin that uses CO_2 and photosynthesis (remember capturing energy from the sun), in exactly the same way that us trees do.

FIG 20 ALGAE AS A SOURCE OF FUEL
(Columns of algae collecting sunlight and converting it into renewable sources of oil with zero carbon footprint.)

This idea is to utilize the capability of algae, that slimy green material often found in still water ponds or water ways to extract carbon dioxide directly from the air, combine it with the energy from the sun and utilize the subsequent oily material produced by the algae as a source of energy. The reason that the process is carbon neutral, or carbon-zero, is because the carbon dioxide that is emitted from the burning of the algae's oil (in effect a biofuel) is the same carbon dioxide that the algae extracted from the air to support its growth and the production of the oil to be used as a fuel. Therefore, there is no net carbon dioxide produced and no increase in greenhouse gas. The company has shown that the process works, and they are now scaling it up to make it economically viable and to demonstrate that they can produce enough oil to have an impact. Using new techniques of genetic engineering and synthetic biology, the company is also working to engineer the algae to make the oil faster, in greater quantities, and using less water. The American military has even used this algae-sourced biofuel as a source of energy to fly its planes and drive some of its ships, at least in support of proof of concept.

There are other sources of new biofuels that may be substitutes for fossil fuels, but most of them still emit a net amount of carbon dioxide into the atmosphere and therefore continue to compound the greenhouse gas problem, albeit at a pace considerably lower than burning traditional fossil fuels. As an interesting aside, and as an example of unintended consequences, there was a push to use corn as an important biofuel. While it worked, it also pushed up the price of a bushel of corn to the point where animal feed became more expensive, as did the price of meat. You might argue that, using hindsight, this wasn't a particularly well thought out initiative.

But even if we can achieve a general source of energy that is close to carbon neutral, it's likely that it won't be enough to solve our climate change problems. Given the level of carbon dioxide and other greenhouse gases in the atmosphere and the extent of global warming that has already occurred, many think that it is not sufficient to merely achieve a neutral carbon footprint. What we really have to do is to actually have a net removal of carbon dioxide from the atmosphere, actually reduce greenhouse gases, not only stop but truly reverse global warming. Here too there are numerous promising research and development projects underway, but unlike the algae project in the works by Synthetic Genomics, they are all still in the research phase short of proof of principle and scale-up.

One process that has been implemented in a small number of efforts is to actually recapture the atmospheric carbon dioxide and store it in giant underground caverns. This seems at first blush a kind of band-aid and even short-term solution since you would expect to sooner or later exceed your ability to store it. But what if rather than recapture and store the CO_2, you could recapture and reuse, not as a biofuel to simply re-release the carbon dioxide, but to reuse for some other purpose. In such a case, you would be achieving a real "net removal" of CO_2 from the atmosphere.

There are many exciting products in development. Some of them recapture CO_2 from the atmosphere and use the CO_2 to make cement and other permanent construction materials, including roadways and highways. Some recapture atmospheric CO_2 and feed them into real greenhouses to be used in plants, thereby increasing agricultural yields. Some are using the CO_2 to produce plastics, and others are attempting to use the CO_2 to produce additional biofuels with zero carbon footprints.

McKinsey & Company, a major international consulting firm, has estimated that by the year 2030, these new CO_2-based products could have values up to one trillion dollars annually and the recapture/reuse of CO_2 could reduce net greenhouse gas emissions by as much as a billion metric tons yearly. Although that's not a huge amount, it has the potential to be significant if we can manage to stop burning fossil fuels and dumping tens of billions of metric tons of CO_2 into the atmosphere each year. If we were able to stop burning fossil fuels and recapture and reuse some of the carbon from the atmosphere, we would not only stop the progression of global warming but even start to reverse it in less than 10 years. Early-stage research, often carried out by young people with fertile and innovative minds, is offering great hope that solutions will be found. It evokes the expression: "Necessity is the mother of invention." These are exciting and critical times for new technologies and the entrepreneurs who drive them. They simply must be encouraged and financed.

So just to review terminology:

1) Burning fossil fuels as a source of energy is "carbon positive" and produces greenhouse gases.
2) Burning algae-derived oil and some other biofuels as a source of energy is "carbon neutral" and has no net impact on greenhouse gases; it neither increases nor decreases them.
3) Capturing and reusing atmospheric CO_2 is "carbon negative," in that it can result in a net decrease of greenhouse gases.

But if you were a cynic, as many of us are, you might also conclude at this point that, talk is cheap. We've been discussing some of these technologies for more than a decade, and yet we're still almost completely dependent on burning fossil fuels, increasing greenhouse gases, and exacerbating the problems of global warming and climate change, not solving them. We, as many others before us, have spelled out the problems in great detail and have also offered statements of what has to be done. We've also offered encouraging words about exciting scientific solutions that are coming along and will be made available "sometime soon." That's all quite encouraging, but less so if "your house is on fire."

You will be surprised to learn, following all that doom and gloom, that the solutions to the climate crisis and global warming are not that complicated. They can be achieved, and they don't require miracles to occur, but they require very strong doses of two essential ingredients: **LEADERSHIP** and **SCIENCE.**

The first part is pretty straightforward, but whoever said that it's easier said than done was spot on.

As a first step on the way to solving the problem of climate change and global warming, we must wrest control of the economy from those who insist on continuing to profit, both directly and indirectly, from global warming, and that of course, is the entire fossil fuel industry and its many appendages. Note for a moment that this is a huge segment of the economy, and for that reason, the issue of taking action against climate change, which should be pretty straightforward, isn't. Too many of these folks are simply too conflicted to make any decisions on the future of energy. If the truth be told, most of them probably understand the dangers of continuing to burn fossil fuels. They are likely too frightened at the thought of losing such a significant part of their livelihoods and are too risk averse to even contemplate the alternatives. In the same vein, most of our politicians are similarly conflicted and fear for the loss of their positions if they take a stance against the fossil fuel industries and their "affiliates," thereby alienating their supporters. These are the same people who oppose rebates for the purchase of electrical vehicles based on a philosophical premise like "that's not the role of governments" rather than their true objective, which is to protect the various arms of the fossil fuel industries.

It's painful to acknowledge that real solutions will be offered only when this whole segment of society loses the ability to profit from fossil fuels and the emission of greenhouses gas, and only then will we have a chance. And who are they? Here are the obvious top 10. There are of course many more. These people aren't necessarily evil, or without concern for our environment. For the most part, they are too vested in the fossil fuel industry and so are too conflicted to be directly involved in major decisions about the future of energy. Decisions are even more complicated because virtually all of us are engaged in multiple ways with these industries. Even if we aren't employed by them, we benefit by their activities, and many of us have interests in those industries in our investments, our retirement plans, and so on.

Here's the list:

1) Oil and gas and coal industries
2) Electricity generated using fossil fuel

3) Financial institutions who support those industries (i.e. all of them)
4) The automotive industry, except for the electric cars initiatives
5) Construction, including steel, aluminum, and cement
6) Agriculture in its entirety
7) Forestry industry involving deforestation
8) Waste management, including landfills
9) Politicians who are beholden to those industries
10) Civil servants who are beholden to politicians and those industries

Now that seems like an impossible task. But it's not. Whole industries have appeared, disappeared, and changed their products in the past. In this case, it has to happen quickly, even abruptly. But note that it's only the first entity listed above, the companies that extract coal, oil and gas from the Earth, that have to disappear. And even they can adapt.

Take for instance Exxon Mobil, the world's second largest public company with posted revenues in 2019 of $255.6 billion. They have already invested heavily in non-fossil fuel energy industries such as biofuels, and they are one of the largest investors in algae-based oils that are carbon neutral. So, it's possible—no, actually—it's essential if those companies are to survive that they will have to dramatically change their "products lines." It'll take time, and major re-investments will have to be made, but it's all doable (see below for the specifics). But mostly it is not a question of what these entities will look like; rather it's a realization that they must shift completely away from the fossil fuel business. It is, however, still an issue of education and of leadership. It's quite simple. Our population must understand that our insistence on continuing to burn fossil fuels as our primary source of energy is nothing less than a catastrophe. Without change, the impact on our lives will continue to deteriorate, potentially even lead to our very extinction. The status quo, which leads to continuing to fill our atmosphere with greenhouse gases, is incompatible with life on Earth as we know it and we must develop the education and the leadership and the will that will bring about the necessary change. Why we are arriving so late to the table is hard to comprehend. It's probably a matter of greed, laziness, or inertia, because these issues are not new. The writing has been on the wall for quite some time now.

Once the leadership is on board, and here, particularly the so-called captains of industry, leaders of major financial institutions, and politicians, the rest is relatively "easy." It is an effort not dissimilar to the American, British, and Canadian decision to massively transform their economies in 1942-1943 in order to win the war against the Nazis in World War II. In 1943 and 1944, 43 percent of the GDP of those countries was applied directly to the war effort. But in this case, the changes would have to be permanent. In terms of financing the initiatives to change, the numbers required to change are not very different from the dollar figures that were just recently allocated to combat the COVID-19 virus, including the trillions of dollars being assigned by the United States, Canada, and other western countries to bolster the economy in these times of change.

SO, HERE IS THE RECIPE OF WHAT WE MUST DO.

1) Ban the extraction and burning of all fossil fuels within a defined period of time, 24-36 months. Note the time frame. It's that serious. Allow for the very substantial compensation (just as we've done in response to COVID-19) of anyone impacted by the transformation in addition to the retraining of those individuals who become unemployed as a result.

2) Invest massively in alternate energy sources that can be built immediately. This has to include major government investments and subsidies in these industries. In the first instance, this would likely be to solar energy and a very large conversion (plus 50 percent) to take place in a time frame of tens of months and not tens of years. Treat it like a Third World War, except here, the enemy is climate change. Nuclear power could follow quickly behind solar, followed in turn by wind and hydrogen power. Remember that there will be

millions and even tens of millions of good, well-paying jobs associated with this conversion.

3) All the massive government subsidies that have been given to the fossil fuel-based industries (both directly and indirectly) should immediately be stopped and applied to those industries providing and developing renewable, carbon-free alternative energy sources.

4) Invest heavily in research into developing new non-fossil fuel based alternative sources, such as nuclear, tidal energy, and renewable carbon neutral or carbon-negative energy sources, including biofuels.

5) Invest heavily and immediately in technologies that have the capacity to reduce atmospheric CO_2 by capturing it and re-utilizing the carbon to produce stable, permanent structures, such as construction cement or permanent structural plastics or other fuels. With a sufficient worldwide effort, it should be possible to reduce atmospheric carbon dioxide from 420 ppm to closer to 350 ppm These actions would not only stop global warming but might even bring temperatures back down to where they were in the 1980s before the sharp increase in temperature occurred.

6) And it goes without saying that ALL countries have to be on board. This is a global issue. We must keep in mind that the greenhouse gases, once they've been released into the atmosphere, have no address.

Note:
It's important to realize that none of this is revolutionary. It's not as if you're being asked to dive into some unknown. There are already communities, albeit few in numbers, who have converted entirely to renewable energy, either solar or wind. There are also whole areas who have relied entirely on hydroelectric energy or nuclear energy and have been off the "fossil fuel grid" for generations.

READILY AVAILABLE SOURCES OF RENEWABLE ENERGY

Solar Energy

Solar energy represents the radiant light and heat that the sun emits that enters our atmosphere. The goal is to harness this energy and use it to replace fossil fuels. Incredibly, in one hour, the sun provides us with the same amount of energy that humans use in an entire year. This is clean, renewable, and non-carbon emitting energy. But, in fact, only 0.4 percent of the energy that we currently use comes directly from the sun. Why is that? Simply put, the power to burn fossil fuels as a source of energy came first and immediately generated enormous profits, which have only continued to grow. Until recently, there was little, if any, impetus for us to figure out how to capture the sun's energy directly. Today, that's changed, and much of the science required to capture the sun's energy without going through the carbon cycle is in place.

But there are two essential problems. The first is that we don't have appropriate battery technologies to store the solar energy for subsequent use when the sun's not shining. The second is the process and cost of changing is very significant given the huge number of different industries that are thriving based on burning fossil fuels as a source of energy. There are the naysayers who will immediately chime in and say that such a conversion is far too expensive and would take decades. The second of those statements is simply not true.

Power Storage

The "beauty" (sic) of fossil fuels, is that they store the sun's energy in the form of carbon-rich energy, deep below the surface of the Earth in the form of coal, oil, and gas, and that we can extract that energy virtually at will. Then we take the energy from deep below the surface of the Earth and store it in coal bins, oil drums, or gas tanks until we were ready to ignite it and get the energy out. The primary problem with solar energy is that we have yet to develop an efficient and cheap process to store the energy for use at a later time. This may be in periods of darkness, heavy cloud cover, or simply when heavy energy surges are required. Rechargeable batteries to store the energy are the obvious answer. Solar energy is currently being stored in batteries, but the battery storage capacities technology has, to date, not kept up with the need. In fact, most of the battery technologies are more than 50 years old. High capacity, relatively low cost, rechargeable batteries are needed and are in development. The good news is that the rates of change in battery technology are spectacular, driven by the electric vehicle initiatives and especially the tremendous advances that the Tesla corporation and their cars has achieved.

Nuclear Power
A few brief words about nuclear power, since it is so often the "Bogeyman" in the room in any discussion of alternative energy sources and climate change. The thought of nuclear power and nuclear power plants conjures up the visions of atomic bombs leveling the Japanese cities of Hiroshima and Nagasaki at the conclusion of World War II. It also reminds us of disasters at three major power plants over the past 40 years: Three-Mile Island (1979), Chernobyl (1986), and Fukushima (2011); and the devastation, in some cases, and fright and panic that their failures brought or at least conjure up.

But the fact is, in the total scheme of things, nuclear power is an excellent alternative to fossil fuels. It's essentially clean in terms of its carbon footprint and the cost of mining uranium, for example as an energy source, is a small fraction of mining the equivalent amount of energy from coal, oil, or gas.

In spite of its reputation, it is also the safest energy source of all. The following table is pretty telling in terms of the relative safety of the different energy sources.

Cumulative Deaths per Terawatt Hour* by Energy Source

Coal	161.0
Oil	36.0
Natural Gas	4.0

Hydroelectricy	1.4
Solar	0.44
Nuclear	0.04

*The Terawatt Hour is a standard unit of energy.

Nuclear power is a proven source of energy, at least for electricity in many countries. France gets 75 percent of its electricity from nuclear power while Hungary, Slovakia, and Ukraine get 50 percent of theirs. There is an issue with spent nuclear waste, but this is thought to be minor with new, small, more efficient nuclear reactors being developed. Bill Gates of Microsoft fame has funded and developed a new nuclear power plant initiative, TerraPower, with new plants that do not use enriched uranium and are not water cooled, so they have some major cost and safety benefits compared to the older, more traditional nuclear power plants.

It seems pretty clear that the main limitations to increased use of nuclear power as a source of energy for all of our needs remains largely emotional and very political. It's hard to imagine, at least at the present time, having a politician run an election campaign based on her/his support for nuclear power. But it is surely an option either for the short or long term. To quote *Wired*, a popular monthly science magazine:

> "Next-Gen Nuclear is Coming—If Society Wants It."………..

Hydroelectric Power

To date, in many parts of the developed world, hydroelectric power is the most important and most widely used renewable energy source, making up about 17 percent of total electricity production. While clean and non-polluting in the traditional sense, hydroelectric power has some major disadvantages. Most of all, it is disruptive. Hydropower facilities usually have major impacts on surrounding land use, homes, and the natural habitats in the large dam area. It can even have an important impact on surrounding watersheds, as water is collected at the dam site. What seems obvious, is that these negative physical and especially social aspects and high building costs make it very unlikely that any significant number of hydroelectric plants will be built in the future, especially in relatively populated areas.

Biofuels or Bioenergy

The notion of burning biofuels or biomass as an energy source as opposed to fossil fuels has received a lot of attention in recent years. Estimates are that as much as 10 percent of the world's total energy demand is currently being supplied through biofuels or biomass. A biofuel is defined as a fuel derived from biomass which can be living or waste material from any living thing (e.g., plants, animals, algae). Biofuels are generally considered to be carbon-neutral because the carbon dioxide that is released into the atmosphere is equal to the amount of carbon dioxide that the biomass absorbed from the atmosphere during its growth over relatively short periods of time. In the case of fossil fuels such as coal, the released carbon dioxide was absorbed from the atmosphere many tens of millions of years ago and more, and so in that case, they are pouring a net amount of carbon into the atmosphere. Biofuels however have some fundamental disadvantages. They require land conversion to produce (i.e. deforestation). They can affect the supply and price of foods for humans and animals by competing for the material, for example corn. While they may be carbon-neutral, their production results in considerable industrial pollution.

Hydrogen

Hydrogen has for many years been considered an exciting source of clean energy. The product of the combustion (i.e., burning) of hydrogen (H_2) with oxygen (O_2) is simply water (H_2O), meaning that hydrogen as a fuel doesn't produce any harmful emissions and therefore can be very environmentally friendly. It is also very efficient. On a weight basis, hydrogen possesses more energy than any other compound, and for this reason, it's hydrogen fuel that NASA uses to power its rockets and its hydrogen fuel cells that provide electricity to the spacecraft. And on top of all that, hydrogen power is entirely renewable. Given that, you might ask, "Why aren't we all driving hydrogen powered cars and deriving all of our electricity from hydrogen fuel cells?" At the present time, the use of hydrogen as a common fuel source is both complex and expensive. In the first instance, it takes a lot of time and effort to separate the hydrogen from other elements, and currently, energy from fossil fuels is used to produce hydrogen fuel, which of course defeats the purpose. Using other renewable energy sources, such as solar or wind energy, to generate the hydrogen fuel would be a much greener choice. A critical limitation to the use of hydrogen as a stable fuel source is that it is highly flammable and relatively difficult to store and therefore to transport.

But the need for clean renewable energy has never been greater, so stay tuned. It might be that with further development, these difficulties might well be overcome, and hydrogen might still one day prevail as an excellent and preferred energy source

Other Power Sources
There are a host of other potential sources of energy, but most of them don't come close to the power or effectiveness of solar and nuclear power. Wind power farms are being built around the world. They too suffer the issue of energy storage, especially on windless days, and too many locals find the wind farms to be intrusive. People have spoken often about the power of the ocean's tides and for generations have asked, why they can't be exploited? Unfortunately, technological advances in this area have been few and far between. Another exciting possibility has been the conversion of the sun's energy, with carbon dioxide from the atmosphere, to make carbon rich energy (e.g. oils) in microbes such as algae. A lot of R&D is being performed on these processes with the hope of developing a real carbon neutral source. We await real proof of principle in terms of cost and scale up.

COSTS TO INNOVATE AND CHANGE
The question of affordability often brings us to a standstill when we think of major changes in the way we do things. For example, a kneejerk response is

always that the cost to convert from oil to electricity or from fossil fuels to solar would be prohibitive. But as we begin to understand the damage that fossil fuels are bringing upon the world, we realize just how poor or really unacceptable an excuse that is for not doing it. Below you'll find the costs of various initiatives as a reminder that when there is a will, there is a way, and that whenever there is a real need, the resources are always found. Then we will paint a fictitious picture of what could happen to a community decimated by the complete loss of its fossil fuel-based economy

COSTS BY THE NUMBERS

$140 trillion	World's annual GDP (gross domestic product)
$6.0 trillion	Stock market losses in the first 10 days of the coronavirus pandemic (2020)
$12.0 trillion	Annual global costs of deleterious impacts of climate change
$5.2 trillion	Annual world subsidies to the fossil fuel industries
$4.7 trillion	Cost to eliminate and replace fossil fuels in the United States
$4.1 trillion	Cost to the United States to win World War II (with the Allies) (based on 2020 dollars)
$4.0 trillion	Global automotive sales (2018)
$2.4 trillion	Annual expenditure on agriculture worldwide
$1.8 trillion	Annual military expenditures worldwide
$28 billion	Annual expenditures on nuclear energy in 2019
$11 billion	Annual expenditures on solar energy

These numbers clearly show that the economic cost to eliminate fossil fuels are large but certainly not beyond the costs that have occurred for other major initiatives or transformations. And as this book was being wrapped up, the American government had spent in excess of 10 trillion dollars battling COVID-19 and shoring up the economy during the pandemic.

The Giving Tree...A Metaphor for Climate Change

FIG 21 SOLAR PANELS VS. SMOKESTACKS

CONVERTING A MYTHICAL ECONOMY

The name of this mythical "self-contained" entity is Kleen. You might want to think of Kleen as an independent country, although it could also be a state, a province, or a region within a country. Coal, oil, and then gas were discovered in Kleen in the early 1900s, and for the last 100-plus years, it has largely been dependent on those energy sectors as its principal economy. Truth be told, it had more than prospered from nature's gift of vast resources of fossil fuels. It has always been considered by others to be a rich community with excellent hospitals, schools, community facilities, and overall relatively low taxes. The name of the community, Kleen, was a bit strange, given the source of its prosperity.

Here are some of its vital statistics prior to "the change," that is the collapse of the pil and gas sector:

Population: (typical age and gender distribution)	5.0 million
Gross Domestic Product (GDP):	$400 billion

GDP from Energy Sector:	$90 billion
Average Annual Investments in the Energy Sector:	$28 billion
Total Workforce:	2.4 million workers
Public Sector Workforce (e.g. health and education)	400,000 workers
Energy Sector Workforce:	250,000 workers
Average Kleen Family Income:	$120.000 gross/year
Average Energy Sector Salary:	$132,500 gross/year
Average Home Cost:	$750,000
Unemployment:	4.5%

It's very clear that the economy and quite frankly the wealth of Kleen was almost entirely dependent on the energy sector, as it had been for many years. Each year, more and more money had been poured into the sector; note the large annual investments. Try as it might over many decades, it had not been able to shed that dependence largely because of the high standards of living, the high salaries, the relatively low population base, and quite frankly the absence of a strong driving force to change.

We ask therefore, what would happen to Kleen if its entire coal, oil, and gas energy sector were to collapse; that is, go to zero over the course of a very short two years as a result of some catastrophic events necessitating the total closing of all fossil fuel-based industries?

Note we are modeling a worldwide collapse of this sector, and alternative energy sources would have to be found for everyone. So from the point of view of access to energy, Kleen would be in the same boat as any other region. It would also be competing with other regions for development in a "new economy."

In the absence of any white knight coming instantly to rescue the economy of Kleen, the economic and social fabric of the region would be devastated. Bankruptcies, home forfeitures, and unemployment would be rampant, and the vital statistics would look grim compared to what they were just a year earlier. Energy for Kleen's own requirements would of course have to be sourced from elsewhere. Here is what the economic indicators would have looked like 18 months after the collapse assuming no mitigation.

Population:	4.7 million
GDP has been reduced to	$200 billion
Energy Sector GDP:	0
Energy Sector Investments:	0
Total Workforce (represents loss of energy sector jobs)	1.7 million)
Public Sector Workforce (including health and education)	320,000 workers
Energy Sector Workforce:	Negligible
Average Kleen Family Income:	$58,500/year
Average Energy Sector Salary:	Negligible
Average Home Cost:	$345,000
Unemployment:	19.5%

These changes would have been devastating for the entire community. Had there been no rescue, it might even be that Kleen would enter total bankruptcy and large sections of it would even disappear. It would have been compared to the disappearance of some of the mining towns in western parts of the United States in the late 1800s and early 1900s. But as it happens, Kleen was to survive, and within a short time, it even began to prosper. A number of smart white knights with deep pockets, a good understanding of energy technologies, and a strong connection to Kleen came forward. And it didn't hurt that a central government agency had introduced strong incentive programs for communities heavily dependent on fossil fuels (either their extraction or their use) in order to transform their economies. The government announced very early on that it would provide matching grants on a dollar-for-dollar basis to companies or individuals who would invest in the development of new industries in these devastated areas. The grants could be for a period of five years with a total maximum government match of $500 billion or $100 billion/year over the five-year period of the redevelopment.

And so it happened that a small group of committed investors with the help of the government established a $500 billion redevelopment fund to create the new Kleen economy. The government provided the $500 billion match. The fund committed to providing grants up to $50 billion over five years (that is $10 billion/year) for the timely development of a new economy for Kleen. And for a number of obvious reasons, aside from the fact that this was their forte, they chose to continue investing in the energy sector, but in place of oil and gas, they chose to develop major assets in solar energy and nuclear power. They hedged their bets and invested roughly half of the funds in each of these new initiatives, new at least for Kleen. The beauty of these investments is that they were in critical technologies not only for Kleen, but indeed for the whole world.

A total of 10 massive new plants were built, four in the solar energy sector and six in nuclear power. In each instance, they brought in very significant partners with deep pockets and excellent track records of bringing technologies to the market. This was not unlike the partnerships that had been established many years earlier, when the oil and gas sectors were being developed with partners such as Shell, Exxon, and BP.

One of the largest solar energy partnerships was established with Elon Musk and the Tesla Corporation given that company's interests and advantages in both solar energy and new battery technology for energy storage.

A nuclear power partnership was established with Bill Gates formerly of Microsoft and the Gates Foundation and a research and development company called TerraPower that Gates and his foundation had formed earlier.

Within 18 months, four plants producing solar energy equipment were up and running and shipping units around the world. Some of the plants were designed to produce small, self-contained units for individual homes, while other mega units were designed for industries and large public areas. In parallel with the solar panel technology, the latest power storage batteries licensed from Tesla were also being manufactured locally and shipped internationally. The initial building of these plants employed a total of 30,000 construction workers, and when fully implemented, each plant employed 12,000 employees working seven days a week and three eight-hour shifts per day. The plants were massive, each occupying in excess of 2.5 million square feet. It took a total of 30 months to go from zero to full capacity. Many lessons were learned from the industrial tech wizard Elon Musk and his experience ramping up the production of Tesla cars in several centres around the world.

The nuclear power plant initiative was much more difficult to implement and more costly. But Kleen and its investors had an advantage. They incorporated and engaged Bill Gates and his nuclear power innovations company TerraPower to give them a major head start. There were major regulatory issues that had to be overcome, and the technology was both newer and more complex. Initially there was some public opposition to the development of nuclear power, but a convincing and effective education program (not just a one-sided lobbying effort) brought most people onboard. It took a full 24 months until two fully functional prototypes were available for testing. The government regulatory process was arduous. There were still a lot of people, especially in government, who were dubious. They generally had all the "traditional" negative vibes and prejudices about nuclear energy. But the new technology proved effective, and the partnership was able to produce small, efficient, and safe reactors for as "little" as $500 million each. The first 100 were in production, and half of them had been shipped fully commissioned before the end of the initial five-year investment period. Within five years, dozens of these "mini" nuclear power plants were being exported from Kleen to sites around the world.

The teams behind the solar energy and the nuclear power initiatives were working on the next generation of their products, and both were spending heavily in the future of their technologies, investing 10 percent of their annual revenues in research. By the fourth year of this initiative, there were 2,500 research jobs in each of the solar energy and nuclear power companies and a further 5,000 individuals engaged in advanced research projects.

In just 60 months following the completion of the initial investment decision, Kleen had undergone a very major "reversal of fortune." Some called the recovery miraculous while others said that it was just smart. Good public policy, a very large infusion of government emergency funding, and excellent leadership within both the investment and industrial sectors made all the difference.

The financial picture of Kleen had improved dramatically.

Population:	4.8 million*
Gross Domestic Product:	$420 billion
GDP from "New" Energy Sector:	$100 billion
Average Annual Investments in New Energy Sector:	$40 billion
Total Workforce:	2.3 million Workers
Public Sector Workforce	400,000 workers
Energy Sector Workforce:	300,000 workers
Average Kleen Family Income:	$132,500
Average Energy Sector Salary:	$121,000
Average Home Price:	$800,000
Unemployment:	5.5%

- During the first 18 months after the collapse of the coal, oil, and gas industries, there was a significant migration away from the region. Within 36 months, that flow of people and talent had started to reverse itself.

Just 72 months following "the crisis," Kleen has returned to its position of prominence in the worldwide energy sector. Its GDP and its "energy sector" GDP are even greater than they were before the collapse of the fossil fuel industry. They are now capturing 6 percent of the world's output of solar panels and solar energy storage units and 8 percent of the world's sales of small efficient nuclear power plants, which were fast becoming the preferred nuclear energy production units. Just as important, the long-term future looked good as Kleen was considered one of the most innovative centers for the development of next generation energy production in both solar and nuclear. Its new industries were clean, innovative, and competitive on a worldwide basis, and once again the salaries in Kleen were some of the highest in the world. You may look at these numbers and respond that they are far too aggressive; that these sorts of turnarounds just don't occur. But this was not a normal time with traditional growth and development timelines. They were at war. Climate change had become a real existential threat, and governments began to take a page from President Roosevelt's response to the threats of World War II and in 1942 established the War Production Board followed in 1943 by the Office of War Mobilization. At that time, entire factories were converted from manufacturing washing machines to producing aircraft or military tanks in less than a year. They needed no less an effort to combat climate change, and they did it.

Note to Reader: This portion of the book (Converting a Mythical Economy) is pure fiction. It is meant as a demonstration of how with extraordinary leadership and appropriate investments, a new energy economy could be established, even in the face of the seemingly devastating closure of an entire fossil fuel industry.

As we come to the end of the story, Let's do a quick summary of the different energy sources, and their pluses and minuses. Let's list the sources starting with those with the highest greenhouse gas emissions going down to the lowest.

Coal
Abundant and cheap
High source of energy
Greatest emitter of greenhouse gas (CO_2)
Emits other toxic material (SO_2, heavy metals)
Enormous health costs to the miners (black lung disease)

Oil
Abundant, cheap, and reliable
Major producer of greenhouse gas (CO_2)
Drilling is expensive and environmentally damaging

Emits other pollutants and greenhouses gases such as methane
Potential spills and inappropriate disposal

Gas
Natural gas is considered a cleaner fossil fuel than oil
It is cheaper, and it burns more efficiently
It gives off about 25 percent less greenhouse gas than oil and about 50 percent less than coal
It is still a major contributor atmospheric CO_2 levels and greenhouse gases
Burning natural gas emits similar toxic pollutants into the air as oil does

Hydroelectric
Clean, renewable; does not pollute water or air
Extremely low carbon footprint once the plant is built
Expensive and dependent on precipitation/water flow
Finite in terms of potential sources
Tremendously disruptive to human and animal habitats

Biomass/Biofuels
Has the potential to be carbon neutral with zero net greenhouses gases
It's renewable, it's cheaper than fossil fuels, it doesn't have to be mined, and it can be produced anywhere
Biofuels have a significant carbon footprint through cultivation
There is also an issue of competition for land space for agriculture and for food; i.e., if you grow corn to be used as a biofuel, are you decreasing food/feed availability or making it more expensive?

Nuclear
Insignificant release of any greenhouse gases
More efficient than fossil fuels
Reliable and low cost
Generates radioactive waste which must be disposed of
Nuclear accidents and serious health effects

Lots of work to make nuclear power plants smaller, more efficient, and safer. Very poor public acceptance at present

Solar

The ultimate "perfect" source of energy

Abundant, renewable, sustainable, clean, available everywhere, low maintenance

But there are some critical problems:

The solar energy is intermittent dependent on the cloud cover and the length of the days (e.g. winter, summer)

Initial infrastructure is expensive, and manufacturing is not especially clean. But the major issue is energy storage, and this technology will bloom when batteries are developed to store the energy for periods when "the sun don't shine"

Wind

Clean energy source with zero carbon footprint

There is no associated pollution of either air or water

Fuel is free, available worldwide, and operating costs are low

Winds are often intermittent and unpredictable

Energy storage issue is the same as with solar energy

There is a resistance to living near "wind farms"

Hydrogen

Hydrogen is the cleanest of all the energy sources

Depending on where the H_2 comes from (e.g., electrolysis of methane), the carbon footprint may be either zero or substantial, but still on the low side

The cost to convert to hydrogen would be expensive

Given the flammable nature of hydrogen, there are serious issues with storage and transport which have to be overcome

Chapter 18

R.I.P. (REST IN PEACE) OR REQUIEM FOR A TREE

FIG 22 THE BOY AND THE FALLEN TREE

In this book, we've attempted to bring you the facts of global warming and climate change. First, from the point of view of a boy representing all humans, then from the point of view of a tree representing all trees and vegetation, and even other forms of organic matter, including that which are now long dead commonly known as fossil fuels. We've described how the combination of population growth, industrialization, and the use of fossil fuels as an energy source have created a perfect storm (pun intended) that has put all of humanity and much of the planet as we know it at tremendous risk.

The notion that this is just a phase and "this too shall pass" is simply not an option. This is not an opinion. It is a statement based on the facts. It's a simple fact that mankind has worked very hard over many years to get us into this mess, and it's pretty clear that we will all have to work even harder to get us out of it

Lao-Tzu, the Chinese philosopher who lived around the start of the common era (50 BCE) and is credited as the founder of the Chinese philosophy, Taoism, is also credited with a quotation that nicely describes where we currently stand in respect to the crises of global warming. *"If you do not change direction, you may end up where you are heading."*

Following this chapter, you'll find an extensive bibliography of references and additional reading material. Often, website information is included that was very helpful in putting together material for the book. In particular though, if you continue to be interested in the subject, please learn more about the organization: Foundation for Climate Restoration (www.foundationforclimaterestoration.org). Their stated mission is to catalyze action to restore the climate by 2050. They maintain an active collaboration of partners throughout the world. They focus quite clearly on three objectives and keep it pretty simple:

1. *Remove the trillion tons of excess CO_2 from the atmosphere*
2. *Restore Arctic ice to prevent catastrophic methane emissions*
3. *Restore life to the world's oceans*

These are all achievable goals if we apply our science and rid ourselves of our most serious fossil fuel addictions. But it's clearly not that easy for each of us to feel the same imperative. It's not as if most of us have a direct, everyday attachment to the excess CO_2 in the atmosphere, to the methane leaching out of the permafrost in the arctic, or to the acidic conditions found in our oceans. Margaret Atwood nicely reminds us in her book *The Testaments*:

> **"You don't believe the sky is falling until a chunk of it falls on you."**

In following up on our story of the boy and the tree, what is very clear is that by the year 2000, the relationship between the boy and the tree deteriorated, and it had done so irreversibly. The breakdown in the relationship probably started as early as the beginning of the second Industrial Revolution in Britain and the advent of the steam engine. In fact, many or even most us, if we put our minds to it and look at the data, would understand that it had developed into a full-blown crisis some time ago. And today, many believe humanity's very existence, or at least our life on Earth as we know it, is in very serious jeopardy. All this is a direct result of our continued dependence on burning fossil fuels. A second critical and unsustainable practice, again driven by population growth, is massive deforestation of the planet for purposes ranging from building cities and highways (urbanization) to the urgent need to grow more food and hence increase agricultural lands and again cut down more trees.

There seems however to be a fundamental difference of opinion, or maybe it's just a lack of common understanding, between scientists who study energy utilization and the environment and those of us who are "merely" users. The scientists who understand the issues and are able to read and follow the data declare that there is an immediate crisis. It will take years to put a halt to and even reverse global warming, and they state emphatically that you simply **must stop burning fossil fuels, and now..** We should have taken these actions decades ago, and even now, it may be too late, but at the very least, we have to try.

Fearful for the economy and their reputations, a large percentage of our industrial, financial, and political leaders respond rigidly. They argue that climate change and global warming are not as bad as they are made out to be. They continue to suggest that the problems may not be manmade, and they warn against the economic devastation that is likely to occur should any serious attempt be made to shut down the fossil fuel industries. Some even today have taken to calling the whole climate change a hoax. At the same time, many do cede to the notion that climate change is a problem, but their support of alternate energy sources (solar, wind, nuclear) and even electric vehicles remains rather tepid.

Note: A Prediction. This will change, and perhaps with a new American administration, it already has. But it will take a lot of kicking and screaming and probably a catastrophic event or two, and probably an overwhelming financial crisis, to really open the eyes of the leaders for them to take the dramatic actions that are required.

In the final analysis, any decision to take substantial action against climate change will not be based on rhetoric from the different camps arguing on the relative merits or the very existence of climate change and global warming. Even the issue of whether the causes are manmade or not will seem less important. It seems obvious that the drivers of change will be purely economic. The global economic impact of climate change is starting to and will continue to "break the bank" and cause financial harm and ruin to many. In an earlier chapter, we reviewed the economic costs of climate change. The wildfires in California and the western states, the hurricanes in southeastern United States, the massive flooding in India and Bangladesh, and the droughts in China and Sub-Saharan Africa represent costs that are increasing annually and are clearly not sustainable.

The social unrest evident in Europe produced by the "illegal" movement of "economic or climate change refugees" from Sub-Saharan Africa amounts to more than one to two million people per year crossing the Mediterranean costing European countries as much as $50 billion per year. Similar movements of people also occur from South and Central America as millions of people attempt to flee north. In China, the number of people fleeing rural agricultural lives on the central plains to seek refuge and greater prosperity in the major cities is astounding. These movements of people occur mostly as a result of climate change and severe droughts year after year in many parts of North Africa and South America. The notion that these economic or political refugees are really "climate change refugees" is now well documented. In 2017, the costs to the Chinese government of extreme weather events, including floods, droughts, heatwaves, and tropical cyclones, exceeded $100 billion. This doesn't include the tremendous social costs to huge segments (200 million plus) of the Chinese population as drought and water shortages force many of them to leave their rural, agrarian lives behind to seek refuge in the new industrial cities sprouting up across China. This is only an example of the real costs of climate change, and the numbers are only growing as the crisis worsens each year.

It doesn't take a lot of imagination to see that this is not sustainable. It will not be long before our financial and political leaders take note of the costs of climate change and begin to insist on change and not in baby steps, as in the Paris Accord, or in notions such as "voluntary carbon taxes," but in real change in energy and financial policies. Many believe, and the evidence for this is increasing each day, that nothing short of the complete ban on burning fossil fuels as a source of energy will be required to make a significant dent in the progression and eventual reversing of global warming.

There have been many "environment-friendly" initiatives that we have been encouraged to take to combat climate change, both as individuals and collectively. Undoubtedly many of you take these very seriously. Here are examples of what we do in favor of our environment.

We:
Recycle	Compost
Ban plastic bags	Ban plastic straws
Keep our houses warmer in summer	Keep our house cooler in winter
Limit meat consumption	Become vegetarian

Become vegans	Rethink packaging
Plant a tree	Plant a million trees
Plant a trillion trees	Walk more than drive
Buy an electric car	Vacation locally (i.e., don't fly)
Use sustainable building materials	Conserve water

These are all laudable initiatives and are no doubt very important. But there is a real danger in not seeing the forest through the trees (pun intended). The largest of these initiatives called the Trillion Trees Initiative has gained a lot of interest, and at the most recent (2020) World Economic Forum in Davos, Switzerland, it was as if signing on was a badge of honor, or the price of admission to the conference. What a fantastic possibility. It almost seemed to be the answer to climate change. Not to be too cynical, but maybe you might think that if you planted a bunch of trees, you wouldn't have to give up driving your gas-guzzling SUV. Certainly, a massive tree planting program is a terrific idea. It will result in the absorption of important amounts of carbon dioxide from the atmosphere. It will provide habitat for animals, and in many cases, it will help to restore fragile ecosystems. But there is a problem or perhaps a warning. The trillion-tree initiative and all other initiatives listed above are terrific, but they must not be a distraction. They must not steal the spotlight from the true primary solution to global warming, and that is to cut emissions and not in a moderate, scaled fashion as suggested in the Paris Accord, but in a manner that matches the real climate crisis we are all facing.

And again, to quote Greta Thunberg speaking at the 2019 Davos World Economic Crisis because she understands what the science is saying.

"I want you to act as if your house is on fire, because it is."

FIG 23 AN EMPASSIONED GRETA THUNBERG SPEAKING TO THE GENERAL ASSEMBLY OF THE UNITED NATIONS

And so, the solution, while not easy or pleasant, is obvious; we must cease all further mining, extracting, and burning of fossil fuels. Remember that billboard we referred to early on? It rings true here.

THE INDUSTRIAL REVOLUTION BROUGHT THE DEVELOPED WORLD 150 YEARS OF UNPRECEDENTED PROSPERITY

GLOBING WARMING IS THE BILL

A BILL THAT HAS NOW COME DUE

The immediate response to such "blasphemy," at least from the politicians and those with vested interest in the broad fossil fuel industry will be, "That's simply not possible." And the simple retort is: "I'm sorry, but the facts say otherwise, and facts don't lie." We know and understand the science of global warming, and we know the essential role that man's use of fossil fuels has played. Furthermore, we have the technologies that are required to replace fossil fuels. What we "simply" need is the political will, and the leadership of politicians and industry, including the investment community to come to the table.

The chapter is titled Rest in Peace (R.I.P.), or in other words Requiem for a Tree, in the belief that we have no choice but to let those dead trees, vegetation, and other organic matter that have languished below the Earth's surface for tens and hundreds of millions of years and more to simply stay there untouched. We must say goodbye to the mining, extraction, and use of all of fossil fuels: coal, oil, and gas. They served us well for so many years, but we didn't pay attention to what they/we were doing to our environment, and today that has produced a real existential threat. And at the same time, we must pay homage and indeed nurture (and plant) those trillion plus trees "who" can help us live in greater environmental peace in the future and even help to reverse global warming.

And so now, the roles have been effectively reversed. For so many years, 400,000 or so to be exact, the boy was well taken care of by the tree. In fact, the boy became ever so dependent on the tree, that by the early 2000s, that dependence started to threaten the boy's very existence. And now, in order to save himself and indeed all of humanity, he must take special care of the tree and especially let those that have fallen rest in peace.

There is a biblical saying from John 8:32: "You will know the truth, and the truth will make you free." That is most pertinent to the existential threat that we have brought upon ourselves through global warming and climate change. We now have the responsibility to ourselves, to our families, and to our communities to be better equipped to know what the truth really is, and that has been an important objective of this book.

And finally, it is most appropriate to end with a quote from arguably the most important scientist of our time, Albert Einstein; a statement that is most relative to our need to understand and spread the truth, and most of all, it is a call to action.

"The world will not be destroyed by those who do evil, but by those who watch and do nothing."

Shel Silverstein's book The Giving Tree, ends with the following final words: "And the tree was happy." But what if we read the book in reverse? What if instead of the boy taking from the tree until the tree is reduced to a stump, instead we find the boy giving to the tree, in essence to nature. At the end of the book, we would see a luscious and thriving tree, and a boy playing in its shade. "And every day the boy would come, and he would gather her leaves, and make them into crowns, and play king of the forest." The boy is our children, our grandchildren and their children. He is once again enjoying the tree because of the decisions and actions actions that we are taking today.

APPENDIX

A HANDBOOK TO HELP YOU IF YOU HAPPEN UPON A DISCUSSION OR, INDEED, IF YOU HAVE TO DO COMBAT WITH A CLIMATE CHANGE DENIER

We assume that since you've gotten this far in the book, you have some sense that global warming is a real issue and that it poses a real and indeed an actual existential threat to all of us. In all likelihood on more than one occasion, you've bumped into a friend, a colleague, or a random person who just doesn't believe it; a climate change denier. This Handbook will provide you with some ammunition to go toe-to-toe with that climate change denier

The first sections will give you the facts (what a novel idea) to support the notion of climate change or global warming, and then at the end, we'll provide you with some tactics to win the argument.

But to start with, who are these people who would deny reality? Who are they, and why do they do what they do?

First and foremost, the major perpetrators of these falsehoods either directly or indirectly represent the fossil fuel industry. Direct members of that industry, political lobbyists, media experts and individuals who have spent the past 30 years sowing doubts about the reality of climate change…when in reality, there are none. And they're no fly-by-night types or one-offs. It's been estimated that the world's five largest oil and gas companies spend a total in excess of $200 million each and every year with the express objective of blocking, delaying, or controlling any public policy initiatives on climate change. Then of course there are the followers, those who follow conspiracy theories as in "global warming is a hoax" perpetrated by the left wing, the progressives, and big government. They are of course encouraged by the large "Trump-following," who regaled in Trump's rejection of climate change and attempt to totally dismantle the Environmental Protection Agency (EPA).

Thankfully, there is hope. In recent years, there has been some substantial resistance or opposition to that lobbying with organizations emerging, such as School Climate Strikes championed by climate change superstar Greta Thunberg, Extinction Rebellion protests, and even national governments declaring climate change emergencies. The opposition has been helped by the media extensive coverage of climate change and the increasing number of extreme weather events. But the lobbying has become even more intense; sometimes subtle and sometimes vicious. Senior elected officials in the United States have even engaged in what has been termed "climate sadism"; for example, when the former president of the United States and senior senators mock and attack young people concerned about their environment and ruthlessly mock then 16-year-old Greta Thunberg, a young girl with Asperger's Syndrome who is simply trying to tell some scientific truths. You simply can't get much lower than that, and really, if you think about it, these people (the climate change deniers) must be getting pretty desperate, and so they should.

Just so that we can better understand these people and more effectively address their issues, let's review the five main reasons why people continue to be climate change deniers.

1) They deny based on science.

They say that climate science is not accurate, that climate change is just part of a natural cycle, that there is too little CO_2 in the atmosphere to make a difference, that climate scientists are fixing the data, that many scientists don't believe in climate change and especially they negate the fact that it is manmade, and the list goes on. All absolute rubbish.

2) They deny based on economics.

A subtle form of climate change denial is that it is too expensive to fix, so really let's try to just ignore it. But that's too expensive for who? In fact, it makes excellent sense to fix the problem and now. The current annual world GDP is about $180 trillion. Setting aside 1 percent of that for up to fifteen years would be sufficient to rid the world of fossil fuels and provide alternative energy sources. That doesn't include the tremendous economic benefits, especially jobs, in introducing the new alternative technologies. Mostly, they are simply conflicted, largely owing their livelihoods to a variety of the fossil fuel industries.

3) They deny based on humanitarian issues.

In some areas of the world, warmer temperatures may appear to be a blessing. As an example, warmer temperatures may increase the growing season in more northern and more southern areas of those respective hemispheres. Farm productivity can be enhanced in some of the more intemperate regions of the Earth. That has to be the absolute height of selective understanding. Say only what you want people to hear and ignore the enormous impact on global warming on agriculture, extreme heat, drought, flooding, and extreme weather such as wildfires or hurricanes.

4) They deny based on politics.

The denial goes something like this. We can't make changes or spend money to combat climate change because other countries are not taking action. That is the height of hypocrisy and the lowest form of leadership or lack thereof.

5) They deny the crisis.

This denial is straightforward. They deny the very existence of climate change. The science isn't true, and even if it were somewhat true, the cost to make any serious correction is too high, so there can't really be a crisis.

FOSSIL FUEL ADDICTION

Another way to look at the issues of climate change denial is to consider the use of fossil fuels as our major source of energy as an **addiction.**

- The "substance" is of course fossil fuels.
- The "users" are all of us who are dependent on fossil fuels for all our energy usage.
- The "dealers" are all the coal, gas, and oil companies that extract the fossil fuel and prepare them for market.
- The "enablers" are all those banks, financial institutions, corporations, investors, and of course politicians who profit from fossil fuels.

The solution, as in all cases, is not complicated. It's also of course not easy.

- In dealing with the actual addiction, you have to remove the "substance" and give the "user" an alternative; and
- The dealers and enablers have to be removed from the picture at least in respect to their profiting from the addictive substance.

But here too there are important and profitable alternatives.

First, it might be worthwhile for you to clarify in your mind some of the terms used in the climate change/global warming discourse. The following definitions are put forth by the Earth Science Communication Team at NASA's Jet Propulsion Laboratory at the California Institute of Technology.

1) Weather refers to the atmospheric conditions that occur locally over a short period of time from minutes to hours to days, as in Mark Twain's famous quote, "If you don't like the weather in New England, just wait a few minutes."
2) Climate refers to the long term regional or even global average of temperature, humidity and rainfall, and other precipitation patterns over seasons, years, or even decades.
3) Global warming, as the term is commonly used, refers to the long term warming of the planet since the early twentieth century and most notably since the late 1970s due to the increase fossil fuel emissions since the industrial revolution.
4) Climate change refers to a broad range of global phenomena created primarily by the burning of fossil fuels, which add heat trapping gases (carbon dioxide being the most important) to the Earth's atmosphere. These include increased temperature trends described by global warming but also the rise of sea levels; ice mass loss in Greenland, Antarctica, the Arctic, and mountain glaciers worldwide; shifts in flower/plant blooming; and extreme weather events.

You'll realize therefore that global warming is really a subset of climate change. There are many other changes that are going on, but it is most likely that the increase in atmospheric temperatures, global warming is the primary driver of those changes. Inasmuch as a lot of the common reporting uses the two terms interchangeably, we've done much the same throughout this book, even though it's not strictly precise.

Now for the 20 questions:

Remember in any case that you're dealing with people who either deny the truth or have succumbed to believing the climate change deniers for whatever reason. For obvious reasons, emotions run high. But if you stick to the simple truths and base your arguments on science, you'll win the day.

Q: How can there be global warming given that this past winter was the coldest ever?

A: Come on now, get serious. Even you know that a couple of cold even freezing days or even weeks doesn't negate the years of warming weather that have actually been measured. Someone once said: "Less cold doesn't mean never cold." Today extremely hot days are 100 times more likely to occur than they did in the years between 1951 to 1980. Those are the real measurements. Remember that global warming causes massive changes in weather, which can explain things like the polar vertex over the Midwest making for days of frigid weather even in the face of global warming. An excellent source of information is the well-respected National Oceanic and Atmospheric Administration, or NOAA for short.

Q: Isn't global warming good for agriculture?

No. It is not. There may be pockets of areas where increased temperatures may expand and enhance growing periods. But there are far more cases where global warming has had very negative impacts on agriculture. Drought in many areas of Africa, central China, and South America have brought crop yields to zero in many areas. In 2010, a major heatwave around Moscow killed 11,000 people and devastated the Russian wheat harvest causing prices to increase worldwide. Flooding along the coasts of Africa and India have negatively impacted crop yields and, in some cases, destroyed agricultural land entirely.

Q: The climate has changed many times in the past. This can't be such a bad thing; it's not like another ice age or something.

A: Yes of course climates have changed in the past. There have been recorded periods of intense heat, and there have been ice ages. But these events occurred over periods of time ranging from 20,000 to 100,000 years. There are physical reasons for those changes in the climate. Sometimes it was the sun burning hotter or colder. Sometimes it was extreme volcanic activity here on Earth. But this time, the current global warming is occurring over a short period of time (50 years), and there is excellent evidence to show that WE ARE TO BLAME. This includes population growth, intense industrialization, increased burning of fossil fuels, and an increase in green house gases, all of which lead directly to global warming. And waiting for some physical event to occur to save us is not a smart idea. We have to act to reverse global warming, and we have to act soon.

Q: But CO_2 is such a minor component of the atmosphere, it can't possibly produce global warming.

A: It is true that CO_2 is only a minor component of our atmosphere (currently 419 ppm or about .4 percent). Nitrogen, oxygen, and argon make up more than 99 percent of the atmosphere. The difference is that none of these gases absorb much of the suns energy (heat) whereas carbon dioxide as well as water vapor and other greenhouse gases such as methane and nitrous oxide do. Here, is a brief account how it works and why it's bad:

The Earth absorbs light (visible) from the sun, and this causes heating. In turn, the Earth's surface and the atmosphere send out infrared radiation back into space and this causes cooling. The greenhouses gases in the atmosphere (CO_2 being the most important) absorb some of the infrared radiation (heat), preventing it from going out back into space and retaining that heat within the atmosphere...hence warming the Earth's atmosphere and producing global warming. The CO_2 is called a greenhouse gas

Atmospheric CO_2 now stands at 419 ppm, the highest in more than 1,000,000 years. Between 1,000,000 years ago and 1900, CO_2 levels ranged between 180 ppm and 300 ppm. The levels have increased from 305 ppm to 416 ppm between 1950 and 2019, which not incidentally coincides with the sharp increase in global warming.

Q: Carbon dioxide must be important since it's critical for all plants to grow?

A: Yes, it is critical for plant growth. The plant takes CO_2 from the air, combines it with water to make certain types of sugars like starches, which are the basis of the plant, and it also gives off O_2, which is important for us to breathe. But as we burn more and more fossil fuels and cut down more and more forests, there develops an imbalance where more CO_2 is being put out into the atmosphere by burning fossil fuels than is taken from the atmosphere by plants and trees to support their growth. That excess of CO_2 in the atmosphere is by far the most common cause of global warming.

Q: How can a little bit of carbon dioxide be so bad for our oceans?

A: The problem is that it's not just a little bit. Annually, there is a net release of 9.3 billion tons of carbon into the atmosphere. That's the difference between the amount of carbon put into the atmosphere and the amount taken out, for example by plants; 2.5 billion tons of that carbon is then dissolved into the ocean's water. Dissolving carbon dioxide into water causes acidification of the oceans, resulting in the destruction of coral reefs and impacting and even killing many other aspects of marine life.

Q: The amount of CO_2 produced by humans is a tiny percent of total CO_2 emissions.

A: That statement is deceiving. The natural cycle (e.g., plant photosynthesis) adds and removes CO_2 from the atmosphere in a good balance. When humans burn coal, oil, and gas in excess, they are only adding CO_2 into the atmosphere without removing any, thereby putting that balance out of sync and causing atmospheric CO_2 to increase.

Q: Volcanoes emit much more carbon dioxide into the atmosphere than humans do.

A: Not true. In fact, on an annual basis, humans send 60 times more carbon dioxide into the atmosphere (as a result of burning fossil fuels) than all volcanoes do in total in an average year.

Q: More people die from the cold than from the heat.

A: Not true. In North America, four times as many people die from the heat than from the cold. And this may be an underestimate since often the cause of death is listed as heart failure, stroke, or respiratory distress when it was really brought on by heat exposure. Most deaths attributed to the cold are the result of poor housing rather than climate change.

Q: Scientists can't even predict next week's weather, how can they measure temperatures 50 or 100 years ago, let alone what it'll be in 100 years?

A: First of all, that's not true. Weather prediction has improved remarkably over the past decade with improved tracking techniques and computer modelling. Yes of course it is not a perfect science, and sometimes they get it wrong. Careful temperature measurements have been made by weather stations around the world for well over a century.

Q: Isn't the ice in Antarctica getting thicker?

A: No. While there has been a report of an increase in sea ice, for reasons that can be explained by changes in water temperature around the land mass, the fact is that satellite measurements show clearly that the loss of land ice is increasing at an alarmingly accelerating rate since the year 2000.

Q: Many scientists don't agree on global warming, and even more don't agree that humans are to blame.

A: That's simply untrue on both accounts. Recent independent polls have shown that the vast, vast majority (98.5 percent) believe in global warming **and** believe that it is humans who are the cause, and yes, primarily by burning too much fossil fuel and cutting down too many forests. The academies of science from at least 80 counties also support this view. And think about it for a moment, global warming correlates in time almost perfectly with the growth of the world population, industrialization, urbanization, the increased burning of fossil fuels, the cutting down of forests worldwide, the increase of greenhouse gases (CO_2 being the most serious) in the atmosphere. There is simply no way that all that can be a coincidence.

Q: How can a few degrees of global warming even hurt us?

A: That might initially seem like a good question, but let's consider it for a moment. How about New York, Toronto, Montreal, Vancouver, San Francisco, London, Paris, and much of coastal Asia, all under water. How about major droughts in important farmlands worldwide? How about the increased need to burn even more fossil fuels to keep our air conditioners on full time? How about sharp decreases in food production in many (but not all) parts of the world. These are already starting to happen, and another 1.5*C will see these disasters become catastrophic unless, that is, we take action.

Q: If we limit carbon emissions, won't we destroy the economy, as in stopping growth, cutting the GDP, and losing jobs?

A: That's simply wrong and dangerously short sighted. The fact is quite the opposite. If we don't limit carbon emissions, we will face enormous economic pressure. It was estimated that in the year 2019, climate changes were responsible directly for 400,000 deaths and cost the world in excess of $1.2 trillion. Just look at the scope of the problem, and you'll realize that the cost to fix the problem will be much lower than the costs associated with ignoring it and using a "business as usual" approach.

Q: Developing a "green economy" is just hoax perpetrated by a bunch of left-wing conspirators.

A: Forget about left-wing, right-wing politics. These are not and should not be political issues. Look at the data. The numbers don't lie. Try to ignore the rhetoric coming from either the left or the right. The science is clear. We're burning more fossil fuel, the Earth is getting warmer, the polar ice caps are melting, severe weather occurs more frequently, and the world is in trouble. Actions will have to be taken to curb global warming no matter what your political stripes are. And by the way, the green economy has the potential to be an incredibly positive economic boon to those who embrace it; many well-paid, high-tech jobs will accrue to those who acknowledge and embrace the new alternative energy sources. We are already seeing tens of thousands of jobs being created in the electric vehicle and solar energy sectors, and it's only the beginning.

Q: Why should we (e.g. United States and Canada) take responsibility for climate change when it is a "global issue"?

A: That's actually a very important question, and there are several answers.

1) Because it is the moral thing to do.
2) Because we (the United States, Canada, Europe, Saudi Arabia, and increasingly China) produce the most carbon dioxide, averaging 17 tons of CO_2 per person as opposed to the world average of less than five tons/person.
3) Because we have the technology and the resources to change and be a beacon for the rest of the world. When we make the appropriate investments to change from a carbon economy to one that doesn't induce global warming, the rest of the world will benefit and will be forced to follow suit.

Q: Any attempt to convert our economy from one based on carbon-based fuels to one based on other energy sources will fail based purely on economic issues.

A: Quite the opposite is true. Converting our economy away from a fossil base will in the final analysis save lives, save money, and perhaps even save humanity from possible extinction. The current world GDP is $180 trillion per year. With only 1 percent of that or $1.8 trillion a year over 15 years, we could convert the economy to a non-carbon economy. We would do away with the $ five trillion a year in annual subsidies that the oil and gas industries currently receive and invest those dollars into alternate energy sources—solar, wind, nuclear, bio, and others.

Q: Animals and plants are very adaptable.

A: That's simply not true on the time scale that we're talking about. Plants and animals evolve over a period of tens of thousands of years and more in a Darwinian fashion. Plants and animals only have some moderate ability to adapt to minor changes in their so-called lifestyles over short periods of time.

Q: Global warming is probably a real thing, but it isn't so bad, so let's wait and make sure that the science is all in and that we truly understand what's going on.

A: This is a stall tactic – a ploy that the climate change denier uses to disarm you as in: "So, okay, maybe you're right, maybe there is such a thing as global warming, but let's be cautious, let's make sure that we get it right." Your combat must be emphatic. No, the science is clear; the evidence is overwhelming, and the longer we wait to take action, the more serious and potentially irreversible the damage will be.

Q: Isn't it too late to do anything about it, even if were to go to a 100 percent clean energy source?

A: That's a tough one to answer with certainty. We have to believe that it's not too late because that likely represents the only future that we have. We've done a lot of damage, but there's no suggestion that we've actually reached any tipping points where the impact of global warming may have become irreversible. In any case, attempting to save the world from a potential "extinction crisis" is surely an imperative under any circumstance. Let's hope that's the case, because if it isn't then we're all doomed. So we have to at least try, and we are increasingly being presented with new sources of clean energy and new ways to actually decrease the carbon dioxide in the atmosphere. This has the potential to not only stop global warming but to reverse it. Unfortunately, many of these technologies are expensive, and too many of us are simply not willing to make the effort, let alone pay the price. There simply isn't a choice.: We have to do it.

The above are only 20 of the most common "untruths" that the deniers use to perpetrate their fraud. There seem to be an almost limitless number of other baseless claims. For example: Mars is warming; polar bear numbers are increasing; Mt. Kilimanjaro's ice loss is due to land use; 500 scientists refute the consensus: they changed the name from global warming to climate change; peer review was corrupted; renewable energy is too expensive; solar cycles cause global warming…

The list goes on, and every one of these statements is blatantly false. But the comment that we hold with the greatest contempt is: "There is no empirical evidence." It sounds so authoritative, but it's a lie. Quite the opposite is true. There is outstanding empirical evidence that global warming and climate change are real, and they are the products of man's activities, primarily the fossil fuels as his primary source of energy.

And finally, to requote Albert Einstein relative to our need to understand and spread the truth, and most of all, call us to action:

"The world will not be destroyed by those who do evil, but by those who watch and do nothing."

SOURCE REFERENCES and ADDITIONAL READING

Evolution of the Problem

Jeanette Winter. *Our House is On Fire, Greta Thunberg's Call to Save the Planet* (2019). Beach Lane Books.

Shel Silverstein. *The Giving Tree* (1964). Harper & Collins.

Mark J. Poznansky. *Saved by Science, The Hope and Promise of Synthetic Biology* (2020). ECW Press. www.savedbyscience.org.

Genetic Literacy Project: Science Not Ideology. https://geneticliteracyproject.org.

Thomas L. Friedman. *Thank You for Being Late, An Optimist's Guide to Thriving in the Age of Acceleration* (2016). Farrar, Straus, and Giroux.

National Environmental Education Foundation (NEEF). www.neefusa.org.

"Predicting Climate Change: Understanding carbon cycle feedbacks to predict climate change." https://www.eurekalert.org/pub_releases/2019-02.

"Plant Evolution and Paleobotany – Rise of Trees." https://sites.google.com/site/paleoplant/narrative/first-forests.

"Age of Man: Enter the Anthropocene." *National Geographic*. www.nationalgeographic.org/article/age-man-enter-anthropocene.

Tom Kompas, Van Ha Pham, Tuon Nhu Che. "The Effects of Climate Change on GDP by Country and the Global Economic Gains from Complying with the Paris Climate Accord" (2018).

Earth's Future

"Extinction." *National Geographic*. www.nationalgeographic.org/encyclopedia/extinction.

Jonathan Safran Foer. *We Are the Weather, Saving the Planet Begins at Breakfast* (2019). Hamish & Hamilton.

Michael Christie. "Greenwood" (2019). McClelland @ Stewart.

"Geological Time Scale; A Timeline for the Geological Science." https://geology.com/time.htm.

"Is There Enough Food for the Future?" Environment Reports: Food Matters (Oct. 7, 2019).

Wilson Stewart and Gar Rothwell. "Paleobotany and the Evolution of Plants" (2010). Cambridge University Press.

Michael Jordan. *The Beauty of Trees* (2012). Quercus.

Jared Diamond. *Collapse* (2006). Viking Press.

Project Drawdown. www.projectdrawdown.org.

"How Trees Changed the World." *Science and Evolution* (Dec. 24, 2007).

PHG Foundation. A Health Policy Think Tank. www.phgfoundation.org.

Ken Gillingham on the Environment and Economics. https://www.youtube.com.

Ken Gillingham. "Carbon Calculus: For deep greenhouse gas emission reductions, a long-term perspective on costs is essential." International Monetary Fund. www.imf.org.

"Live Science: Chart of Geological Time." www.livescience.com.

"Health Benefits far outweigh the costs of meeting climate change goals." World Health Organization. https://www.who.net/news-room/detail/05-12-2018.

"Carbon Footprint Factsheet." Centre for Sustainable Systems (Aug. 28, 2019). www.css.umich.edu.

State of the Planet, Earth Institute Columbia University. https://blogs.ei.columbia.edu.

"Junk Science Week: Science Is on the Verge of a Nervous Breakdown." https://business.financialpost.com.

Global Warming and Climate Change are Real

Jeff Goodell. *The Water Will Come: Rising Seas, Sinking Cities, and the Remaking of the Civilized World* (2017). Back Bay Books.

National Environmental Education Foundation (NEEF). www.neefusa.org.

"Capturing Carbon's Potential." State of the Planet Earth Institute, Columbia University. www.earth.columbia.edu.

"El Nino and Global Warming, What's the Connection?" State of the Planet, Earth Institute, Columbia University.

Kevin Doxzen. "Make a Sustainable choice: Buy GMO Food." San Francisco Chronicle (April 6, 2018). www.sfchronicle.com/opinion.

"Climate Change Facts: Effect on the Economy." https://www.thebalance.com/economic-impact-of-climate-change-3305682.

"Climate Change: How Do We Know?"...Just Google it.

"Climate Change: Vital Signs of the Planet." NASA.

Hellen Kollias. "Are GMOs bad for your health? If you're asking this question, you're probably missing the point." www.precisionnutrition.com.

Christine Nunez. "Is Global Warming Real? Scientific consensus is overwhelming: The planet is getting warmer and human are behind it." *National Geographic* (Jan. 31, 2019). www.nationalgeographic.com.

"Global Warming is contributing to extreme weather events" (Jan. 8, 2020). https://sites.nationalacademies.org.

Julian Cribb. *The Coming Famine: The Global Food Crisis and What We Can Do to Avoid It.* University of California Press.

"Migration and Climate Change." International Organization for Migration. www.iom.int.

Whit Bronaugh. "North American Forests in the Age of Man." www.americanforests.org/magazine.

"What's driving Deforestation? Union of Concerned Scientists." Science for a Healthy Planet and Safer World. https://www.ucsusa.org/global-warming.

"How many trees would I have to plant to solve Global Warming?" Earth Science Stack Exchange (Aug. 28, 2019).

"Americans are planting...Trees of Strength." NC State University, College of Agriculture & Life Science. https://projects.ncsu.edu/project/treesofstrength/benefits.htm.

Nicholas Kristof. "Food Doesn't Grow Here Anymore. That's Why I Would Send My Son North." *The New York Times* (Jun. 5, 2019).

"Population: A Problem or Not?" https://populationmatters.org/mythbusting.

"Climate Change: How Do We Know?" National Aeronautics & Space Administration: Global Climate Change.

Diana Kwon. "How to Debate a Science Denier." *Scientific American* (Jun. 25, 2019).

Kimberly Amadeo. "Climate Change Facts: Effect on the Economy" (Jun. 25, 2019). https://www.thebalance.com/economic-impact.

Chelsea Harvey. "CO2 Emissions Reached an All-Time High in 2018." *Scientific American* (Dec. 6, 2018).

"Climate Change and Agriculture." Wikipedia.

Renee Cho. "El Nino and Global Warming—What's the connection?" State of the Planet, Earth Institute/Columbia University (Feb. 2, 2016).

Valenti Rull. "The Deforestation of Easter Island." Biological Reviews, Cambridge Philosophical Society (Oct. 10, 2019).

Karn Vohra et. al. "Global Mortality from Outdoor Fine Particle Pollution Generated by Fossil Fuel Combustion." Environmental Research (Feb. 9, 2021).

"Global Agriculture towards 2050." How to Feed the World 2050: High Level Expert Forum (Oct. 13, 2019).

"25 years of GMO crops: Economic, environmental and human health benefits" (Apr. 5, 2018). https://geneticliteracyproject.prg.

"Is Global Warming Real? Scientific consensus is overwhelming: The planet is getting warmer and humans are behind it." https://www.nationalgeographic.com/environment/global-warming-real.

James Temple. "'A Trillion Trees' is a great idea—that could become a dangerous climate distraction." MIT Technology Review. www.technologyreview.com.

"The Effects of Climate Change on GDP by Country and the Global Economic Gains from Complying with the Paris Climate Accord." https://agupubs.onlinelibrary.wiley.com/doi/full/10.1029/2018.

"How many trees would I have to plant to solve Global Warming?" Earth Science Stack Exchange (Aug. 28, 2019). www.earthscience.stackexchange.com.

There Are Solutions

"Moving the United States toward a 100% clean energy economy." https://www.beyondcarbon.org.

Mark J. Poznansky. *Saved by Science, The Hope and Promise of Synthetic Biology* (2020). ECW Press. www.savedbyscience.org.

Amy Webb and Andrew Hessel. *The Genesis Machine, Our Quest to Rewrite Life in the Age of Synthetic Biology.* Public Affairs, New York (2022).

Foundation for Climate Restoration. www.foundationforclimaterestoration.org.

"2050: How Earth Survived." *Time Magazine* (Sept. 23, 2019).

"Planting 1.2 Trillion Trees Could Cancel Out a Decade of CO_2 Emissions." Yale School of Forestry and Environmental Studies (Jun. 2019).

"How Much Energy Does the Sun Produce? (And Other Fun Facts)." https://www.solarpowerrocks.com/solar-basics/3-reasons-the-sun.

Yale Environment 360, Planting 1.2 Trillion Trees. www.e360yale.edu.

Kate Whiting. "How soon will we be eating lab-grown meat?" World Economic Forum (Oct. 16, 2020).

Bill Gates. *How to Avoid a Climate Disaster, The Solutions We Have and the Breakthroughs We Need.* Alfred A Knopf (2021).

Renee Cho. "The State of Nuclear Energy Today—and What Lies Ahead." State of the Planet. Earth Institute, Columbia University (Nov. 23, 2020).

"Nuclear Power Safety Concerns." Council on Foreign Relations.

"Climate Change Feedback." Wikipedia.

Elizabeth Kolbert. "Creating a Better Leaf: Could tinkering with photosynthesis prevent a global food crisis?" The Control of Nature (Dec. 13, 2021).

"Greta Thunberg, Person of the Year." *Time Magazine* (Dec. 23, 2019).

"Capturing Carbon's Potential." State of the Planet Earth Institute, Columbia University (May 29, 2019). https://blogs.ei.columbia.edu.

"American are planting…Trees of Strength." NC State University, College of Agriculture & Life Science. https://projects.ncsu.edu/project/treesofstrength/benefits.htm.

"Moving the United States toward a 100% Clean Energy Economy Beyond Carbon." Bloomberg Philanthropies. https://www.beyondcarbon.org.

Angel Dewan and Rachel Ramirez. "UN report on climate change crisis confirms the world already has solutions—but politics are getting in the way." CNN Report (Apr. 4, 2022). https://www.cnn.com/2022/04/04/world/un-ipcc-climate-report-mitigation-fossil-fuels/index.html.

Genetic Literacy Project: Science Not Ideology. www.geneticliteracyproject.org.

"Carbon Footprint Factsheet." Centre for Sustainable Systems (Aug. 28, 2018). www.css.umich.edu.

"Moving the United States toward a 100% clean energy economy beyond carbon." Bloomberg Philanthropies. (https://www.beyondcarbon.org)

"Synthetic Biology pulls carbon dioxide out of the atmosphere." Chemical and Engineering News: News of the Week (Nov. 17, 2016).

Peter Hergersberg. "Biologist discusses a synthetic metabolic pathway that fixes carbon dioxide and synthetic biology" (Nov. 25, 2016). https://phys.org/news/2016-11.

"Guidelines for the Appropriate Risk Governance of Synthetic Biology. International Risk Governance Council World Resources Institute. https://www.wri.org.blog.

Capturing Carbon's Potential: These Companies are Turning CO2 into Profits
State of the Planet, Earth Institute/Columbia University (May 29, 2019). https://blogs.ei.columbia.edu

Craig Venter. "Bugs Might Save the World." *New York Times Magazine* (Jun. 3, 2012). www.nytimes.com/2012/06/03/magazine.

Max-Planck-Gesellschaft. "Using Synthetic Photosynthesis to Combat Climate Change" (Nov. 21, 2016). www.mpg.de/10834045.

Printed in the USA
CPSIA information can be obtained
at www.ICGtesting.com
LVHW011738070923
756226LV00002B/18